SpringerBriefs in Applied Sciences and Technology

W0037283

For further volumes:
http://www.springer.com/series/8884

Russell Wanhill · Simon Barter

Fatigue of Beta Processed and Beta Heat-treated Titanium Alloys

 Springer

Russell Wanhill
National Aerospace Laboratory NLR
Aerospace Vehicles Division
PO Box 153
8300 AD, Emmeloord
The Netherlands
e-mail: wanhill@nlr.nl

Simon Barter
Air Vehicles Division
DSTO Defence Science
 and Technology Organisation
506 Lorimer St, Fishermans Bend
Melbourne, VIC 3207
Australia
e-mail: simon.barter@dsto.defence.gov.au

ISSN 2191-530X
ISBN 978-94-007-2523-2
DOI 10.1007/978-94-007-2524-9
Springer Dordrecht Heidelberg London New York

e-ISSN 2191-5318
e-ISBN 978-94-007-2524-9

Library of Congress Control Number: 2011937172

Printed on acid-free paper

Springer is part of Springer Science+Business Media (www.springer.com)

Abstract

This Springer Brief reviews most of the available literature on the fatigue properties of beta (β) annealed Ti-6Al-4V and titanium alloys with similar microstructures. The emphasis is on beta processed and beta heat-treated titanium alloys, because beta annealed Ti-6Al-4V Extra Low Interstitial thick plate has been selected for highly loaded and fatigue critical structures in advanced high-performance aircraft that are currently intended to enter service with several Air Forces around the world. The main topics in this review are the fatigue initiation mechanisms, fatigue initiation lives, and short-to-long (or small-to-large) fatigue crack growth in beta processed and beta heat-treated titanium alloys. However, some comparisons are made with alloys having different microstructures, in particular conventionally alpha + beta ($\alpha + \beta$) processed and heat-treated Ti-6Al-4V.

Keywords

Titanium alloys · Fatigue crack growth · Damage tolerance · Durability

Contents

Abbreviations

a	Crack size
CA	Constant amplitude
CMU	Controlling Microstructural Unit
DT&D	Damage Tolerance and Durability
EBA	Effective Block Approach
EIFS	Equivalent Initial Flaw Size
ELI	Extra Low Interstitial
EPFM	Elastic-Plastic Fracture Mechanics
EPS	Equivalent Pre-crack Size
FASTRAN	Short crack growth model
HCF	High-Cycle Fatigue
HID	High Interstitial Defect
ISY	Intermediate Scale Yielding
K_t	Stress concentration factor
LAD	Low Alloy Defect
LCF	Low-Cycle Fatigue
LEFM	Linear Elastic Fracture Mechanics
LSY	Large Scale Yielding
M	Microstructural unit size
N	Number of cycles
N_i	Number of cycles (life) to fatigue crack initiation
N_{lc}	Number of cycles (life) during long (large) fatigue crack growth
N_{sc}	Number of cycles (life) during short/small fatigue crack growth
N_t	Total number of cycles (fatigue life)
NDI	Non-Destructive Inspection
OEM	Original Equipment Manufacturer
QF	Quantitative Fractography
r_p	Crack tip plastic zone size
R	Stress ratio, S_{min}/S_{max}
S	Stress
S_{max}	Maximum (fatigue) stress

S_{min}	Minimum (fatigue) stress
SSY	Small Scale Yielding
STA	($\alpha + \beta$) Solution Treated and Aged
Y	Geometric factor in LEFM description of cracks
VA	Variable Amplitude
α	Titanium alloy phase with hexagonal close packed (hcp) crystal structure
β	Titanium alloy phase with body centred cubic (bcc) crystal structure
ε	Strain
$\Delta\varepsilon$	Fatigue strain range
ΔK	Stress intensity factor range
ΔK_{th}	Fatigue crack growth threshold for long (large) cracks
ΔS	Fatigue stress range
ΔS_e	Fatigue stress range endurance limit
σ_y^c	Cyclic yield stress

Chapter 1
Introduction

Beta annealed Ti-6Al-4V Extra Low Interstitial (ELI) titanium alloy has a chemical composition and manufacturing process intended to optimise its fatigue and fracture properties, notably in the thick sections required for large primary aircraft structures. This alloy, in thick plate form, has been selected for the main wing-carry-through bulkheads and other fatigue critical structures, including the vertical tail stubs, of advanced military aircraft that are currently intended to enter service with several Air Forces around the world, including the Royal Australian Air Force (RAAF) and Royal Netherlands Air Force (RNLAF).

However, publically available data on the fatigue and fracture properties of beta annealed Ti-6Al-4V are limited. This is particularly the case for the kinds of data required for (a) independent fatigue life assessments that conform to the Damage Tolerance and Durability (DT&D) requirements used by the aircraft manufacturer (Original Equipment Manufacturer, OEM); and (b) reassessments to be made as service experience is obtained. This latter point is most important, for the following reasons:

(1) Service fatigue load histories can and do vary significantly from the design assumptions and the load histories applied during the original full-scale DT&D testing of the airframe.
(2) Advanced military aircraft structures are highly efficient designs that experience relatively high stresses. This means that fatigue issues can arise at features such as shallow radii at the junction of flanges, webs and stiffeners, as well as at holes and tight radii. As a consequence, there are usually many areas that need to be assessed for their fatigue lives, and many locations at which cracking may occur in service, but not necessarily during the original full-scale testing.

In the light of the limited or even non-existent data for independent fatigue life assessments and reassessments of beta annealed Ti-6Al-4V components, the DSTO and NLR set up a joint programme of testing, fatigue life modelling and

R. Wanhill and S. Barter, *Fatigue of Beta Processed and Beta Heat-treated Titanium Alloys*, SpringerBriefs in Applied Sciences and Technology, DOI: 10.1007/978-94-007-2524-9_1, © The Author(s) 2012

Table 1.1 Survey of fatigue life assessment methods in the DSTO—NLR joint programme

Fatigue life assessment methods

- *Strain—Life* (ε—N)
    - ○ Strain—life equation, unnotched data, R = −1
    - ○ Cyclic stress—strain curve analysis
    - ○ Rainflow cycle counting (closed hysteresis loops)
    - ○ Stress—strain at critical location (notch analysis)
    - ○ Mean stress effects (R) via equivalent strain equations, leading to equivalent strain amplitudes
    - ○ Damage accumulation rule
- *Damage Tolerance and Durability* (*DT&D*)
    - ○ Specified equivalent initial flaw sizes (EIFS) based on Non-Destructive Inspection (NDI) capabilities
    - ○ Back-extrapolation of long crack growth data to derive short crack growth
    - ○ LEFM long crack growth models (non-interaction, yield zone, crack opening, strip yield) to derive variable amplitude (VA) crack growth from constant amplitude (CA) data
    - ○ Possible use of crack opening model for short cracks (FASTRAN); differences in long and short crack thresholds need to be included
    - ○ Mainly deterministic: stochastic approach becoming accepted
- *DSTO Flight Block Spectrum Loading* (*Effective Block Approach, EBA*)
    - ○ Actual initial discontinuity/flaw sizes and their equivalent pre-crack sizes (EPS)
    - ○ Actual short-to-long crack growth data using marker loads and Quantitative Fractography (QF)
    - ○ Data compilations to establish empirical relationships describing crack growth behaviour
    - ○ Deterministic ("upper bound") estimates of *lead crack* growth (Molent et al. 2011)
    - ○ Scatter factors

fatigue crack growth analysis for this material. This programme takes account of the conventional DT&D requirements and also an innovative fatigue lifting approach developed by the DSTO. This approach is based on years of inspection and analysis of fatigue cracks in many airframe components and specimens, and is an important additional method of determining fatigue lives for aircraft in the RAAF fleet (Molent et al. 2011).

1.1 Survey of Fatigue Life Assessment Methods

Three fatigue life assessment methods are considered in the DSTO—NLR joint programme. These methods are surveyed and summarized in Table 1.1. Each requires specific kinds of fatigue data and there are major and fundamental differences between them:

(1) The Strain—Life method does not consider initial cracks and does not involve crack growth analyses. However, the DT&D and DSTO-EBA methods are based entirely on crack growth analyses and assume that fatigue cracking begins soon after an aircraft enters service.

(2) The DT&D method specifies a "standard" set of initial crack/flaw sizes based on Non-Destructive Inspection capabilities, while the DSTO method uses initial crack/flaw sizes representative of small, fatigue-initiating discontinuities in the materials and structural components used in the aircraft.

(3) In the DT&D method the important period of early (short) crack growth is estimated from back-extrapolation of long crack growth data. However, in the DSTO method the crack growth lives are estimated from actual data for short-to-long cracks growing from the fatigue-initiating discontinuities.

In view of the joint programme it was considered essential to review what is available in the open literature about the fatigue properties of β annealed Ti-6Al-4V, and also the fatigue properties of titanium alloys with similar microstructures. This is the main purpose of the present publication.

Reference

L. Molent, S.A. Barter, R.J.H. Wanhill, The lead crack fatigue lifing framework. Int. J. Fatigue **33**, 323–331 (2011)

Chapter 2
Metallurgy and Microstructure

2.1 The General Metallurgy of Titanium Alloys

Unalloyed titanium has two allotropic forms. The low temperature form, α, exists as an hexagonal-close-packed (hcp) crystal structure up to 882°C, above which it transforms to β, which has a body-centred-cubic (bcc) crystal structure.

The alloying behaviour of elements with titanium is defined by their effects on α and β. Element additions that increase or maintain the temperature range of stability of the α phase are called α-stabilizers. The most important of these are aluminium, tin and zirconium. Element additions that stabilize the β phase are called β-stabilizers. These include molybdenum, vanadium and iron.

There are also important impurity elements, namely oxygen, hydrogen, nitrogen and carbon. Oxygen and hydrogen are the two most important impurities: oxygen is an α-stabilizer and hydrogen is a β-stabilizer. These four elements are also referred to as interstitial elements. This is because their atomic sizes are much less than those of the metallic alloying elements and they fit in the spaces (interstices) between the crystallographic positions of the metal atoms in the α and β phases.

Titanium alloys can be classified in four categories:

(1) α alloys	Examples are commercially pure grades of Ti, containing well-defined amounts of oxygen, and Ti-2.5Cu and Ti-5Al-2.5Sn.
(2) Near-α alloys	These contain only a small amount of β phase. They are heat-treatable and stronger than α alloys. Early examples are Ti-6Al-2Sn-4Zr-2Mo and Ti-8Al-1Mo-1V. More complex alloys have been developed for improved creep resistance. These include TiAlZrMoSiFe and TiAlZrSnNb(Mo,Si) alloys.
(3) α–β alloys	These contain limited amounts of β-stabilizers, the majority of which cannot strengthen the α phase. Hence α-stabilizers are also added. The mechanical properties depend on the relative

R. Wanhill and S. Barter, *Fatigue of Beta Processed and Beta Heat-treated Titanium Alloys*, SpringerBriefs in Applied Sciences and Technology, DOI: 10.1007/978-94-007-2524-9_2, © The Author(s) 2012

amounts and distribution of the α and β phases. These variables are controlled by processing and heat treatment. Examples are Ti-6Al-4V and Ti-6Al-2Sn-4Zr-6Mo.

(4) β alloys   These have sufficiently high β-stabilizer contents that commercially useful microstructures are predominantly β phase. They have been developed mainly because of excellent formability (e.g. cold-rolling) and very good response to heat treatment. Examples are Ti-15Mo-3Nb-3Al-0.2Si and Ti-10V-2Fe-3Al.

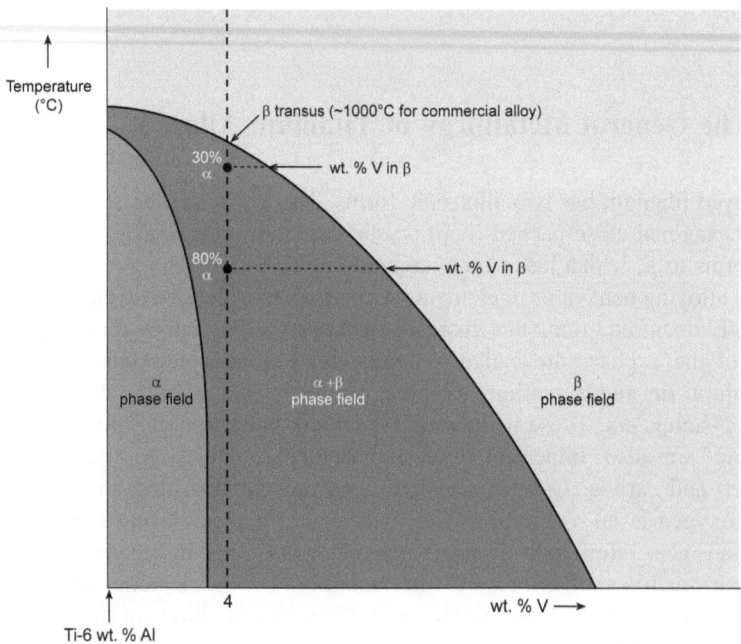

Fig. 2.1 Pseudo-binary equilibrium phase diagram (schematic) for Ti-6Al-4V. The relative amounts of α at the two indicated temperatures are derived from the metallurgical phase diagram *lever rule*

2.2 Conventional Thermomechanical Processing of Alloys Like Ti-6Al-4V

The most important parameter in thermomechanical processing and heat treatment of α–β titanium alloys is the β transus temperature. Figure 2.1 shows a schematic phase diagram for Ti-6Al-4V. Above the β transus the alloy is 100% β and is relatively easily worked. But β working has, or can have, adverse effects on some mechanical properties, and so can β heat treatment, see Sect. 2.3. Consequently, even though initial working may be in the β phase field, final working is conventionally in the $(\alpha + \beta)$ phase field and any subsequent conventional heat treatments are also in the $(\alpha + \beta)$ phase field.

However, this is not the whole story, since the mechanical properties of α–β alloys depend on the relative amounts and distribution of the α and β phases. These variables depend strongly on the thermomechanical processing and heat treatment temperatures in the $(\alpha + \beta)$ phase field. For example, Fig. 2.1 indicates the percentage of primary α in Ti-6Al-4V at two temperatures. At the higher temperature there is less *primary* α and more β, which, however, contains less vanadium. Since vanadium is a β-stabilizer, this means that the higher temperature β more readily transforms to α during cooling after processing, and also during any subsequent heat treatment at relatively low temperatures, e.g. annealing at 700–750°C.

When β transforms to α, the morphology of the *transformation* α is very different to that of *primary* α. Figure 2.2 illustrates this schematically: prior β grains have transformed to co-oriented α lamellae (*transformation* α) separated by ribs of retained β. The crystallographic significance of this transformation is discussed in Sect. 2.4, but here it is worth noting that manufacturers impose property-driven requirements on the relative amounts of primary α and transformed β in conventionally processed Ti-6Al-4V. An excellent overview is provided by Wedden and Liard (1973).

Fig. 2.2 Illustration of an α–β titanium alloy duplex microstructure, showing more or less equiaxed grains of primary α and transformed β. The transformation of β results in co-oriented α lamellae (transformation α) separated by "ribs" of retained β

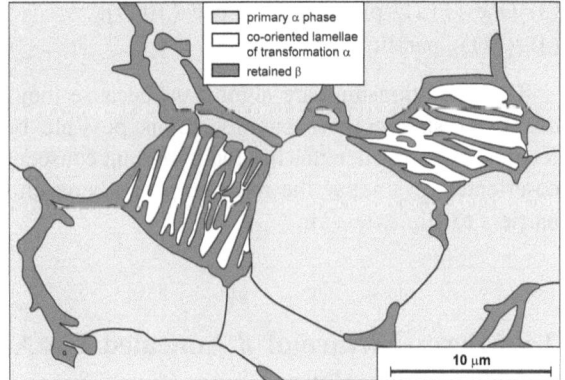

primary α phase
co-oriented lamellae
of transformation α
retained β

10 μm

2.3 β Processing and/or β Heat-treatment

The use of β processing and/or β heat-treatment for α–β titanium alloys is generally regarded as inadvisable. Forging above the β transus is more difficult to control and can lead to coarse-grained microstructures detrimental to alloy strength, ductility and fatigue properties, as can β heat-treatment (Coyne 1970;

Green and Minton 1970; Donachie 1982, 2000; Terlinde et al. 2003; Wagner 1997; Evans 1999).

However, it has long been recognized that β annealing of Ti-6Al-4V, although it significantly reduces the strength, is beneficial to the DT&D-related properties of long/large fatigue crack growth, fracture toughness and stress corrosion cracking resistance (Paton et al. 1976). This recognition has recently led to the introduction of β annealed Ti-6Al-4V thick plate for large primary structures in advanced aircraft, as mentioned in Chap. 1.

It is important to note here that β annealing has become an accepted heat-treatment for optimizing the creep resistance of some near-α alloys used in aeroengine (gas turbine) compressors (Peters et al. 2003). The ambient temperature fatigue crack growth properties of these alloys provide data to be discussed and compared with similar data for β annealed Ti-6Al-4V in Chaps. 5 and 6.

2.4 Crystallography of the $\beta \rightarrow \alpha$ Phase Transition

The transformation of metastable β phase to stable α phase obeys—approximately—the following crystallographic relationships for each transforming β grain (Newkirk and Geisler 1953; Williams et al. 1954):

(1) One $\{110\}_\beta$ plane parallel to $\{0002\}_\alpha$.
(2) One $\{112\}_\beta$ plane parallel to one $\{10\bar{1}0\}_\alpha$.
(3) One $\{112\}_\beta$ plane at 9° to one $\{10\bar{1}0\}_\alpha$.
(4) $\langle 111 \rangle_\beta$ parallel to $\langle 11\bar{2}0 \rangle_\alpha$.

These relationships are significant because they show that slip compatibility, due to crystallographic parallelism, is possible between transformation α and retained β. In practice this has the important consequence that when slip initiates in co-oriented α lamellae the retained β ribs between the lamellae are not effective barriers to slip extension.

2.5 Microstructure of β Annealed Ti-6Al-4V ELI Thick Plate

Figure 2.3 gives an example of the microstructure near the mid-thickness of a thick plate (125 mm) of β annealed Ti-6Al-4V ELI acquired for the DSTO–NLR joint programme. The β annealing heat treatment resulted in a fully lamellar microstructure consisting of large colonies of co-oriented α lamellae within the prior β grains, which are delineated by grain boundary α.

Figure 2.3 also shows that the aligned α platelets are separated by thin ribs of retained β. As discussed in Sect. 2.4, these ribs are likely to be ineffective barriers

Fig. 2.3 Example microstructure (Kroll's etch) of the β annealed Ti-6Al-4V ELI thick plate acquired for the DSTO–NLR joint programme. This example shows large colonies of co-oriented α lamellae within prior β grains, with grain boundary α delineating the prior β grains. The prior β grain sizes in the plate were found to range from about 0.2 to 2 mm, with a mean of 1.2 mm

to slip. However, the colonies themselves have differing crystallographic orientations and would therefore be expected, like the prior β grain boundaries, to act as barriers to slip. In fact, the boundaries between the colonies and the prior β grains are not always effective slip barriers, as will be shown in Sect. 3.2.

References

J.E. Coyne, The beta forging of titanium alloys, in *The Science, Technology and Application of Titanium*, ed. by R.I. Jaffee, N.E. Promisel (Pergamon Press, London, 1970), pp. 97–110

M.J. Donachie Jr., Introduction to titanium and titanium alloys, in *Titanium and Titanium Alloys Source Book*, ed. by M.J. Donachie Jr. (ASM International, Metals Park, 1982), p. 3

M.J. Donachie Jr., *Titanium: A Technical Guide*, 2nd edn. (ASM International, Materials Park, 2000), pp. 34–37

W.J. Evans, Microstructure and the development of fatigue cracks at notches. Mater. Sci. Eng. A **A263**, 160–175 (1999)

T.E. Green, C.D.T. Minton, The effect of beta processing on properties of titanium alloys, in *The Science, Technology and Application of Titanium*, ed. by R.I. Jaffee, N.E. Promisel (Pergamon Press, London, 1970), pp. 111–119

J.B. Newkirk, A.H. Geisler, Crystallographic aspects of the beta to alpha transformation in titanium. Acta. Metall. **1**, 370–374 (1953)

N.E. Paton, J.C. Williams, J.C. Chesnutt, A.W. Thompson, The effects of microstructure on the fatigue and fracture of commercial titanium alloys, AGARD Conference Proceedings No. 185, Advisory Group for Aerospace Research and Development, Neuilly-sur-Seine, 1976, pp. 4-1-4-14

M. Peters, J. Hemptenmaker, J. Kumpfet, C. Leyens, Structure and properties of titanium and titanium alloys, in *Titanium and Titanium Alloys, Fundamentals and Applications*, ed. by C. Leyens, M. Peters (Wiley-VCH GmbH & Co. KGaA, Weinheim, 2003), pp. 1–36

G. Terlinde, T. Witulski, G. Fischer, Forging of titanium, in *Titanium and Titanium Alloys, Fundamentals and Applications*, ed. by C. Leyens, M. Peters (Wiley-VCH GmbH & Co. KGaA, Weinheim, 2003), pp. 289–304

L. Wagner, Fatigue life behavior, in *ASM Handbook Volume 19 Fatigue and Fracture, Second Printing*, ed. by S.R. Lampman et al. (ASM International, Materials Park, 1997), pp. 837–845

P.R. Wedden, F. Liard, Design and development support for critical helicopter applications in Ti-6Al-4V alloy, in *Titanium Science and Technology*, ed. by R.I. Jaffee, H.M. Burte (Plenum Press, New York, 1973), pp. 69–80

A.J. Williams, R.W. Cahn, C.S. Barrett, The crystallography of the β–α transformation in titanium. Acta. Metall. **2**, 117–128 (1954)

Chapter 3
Fatigue Initiation Sites

3.1 Microstructural Initiation Sites in Near-α and α–β Alloys: Literature

Figure 3.1 illustrates schematically the reported fatigue crack initiation sites for the three main microstructural categories of near-α and α–β alloys (Wells and Sullivan 1969; Stubbington and Bowen 1974; Eylon and Pierce 1976; Eylon and Hall 1977; Postans and Jeal 1977; Ruppen et al. 1979; Bania et al. 1982; Bolingbroke and King 1986; Wojcik et al. 1988; Dowson et al. 1992; Evans and Bache 1994; Demulsant and Mendez 1995; Lütjering et al. 1996; Wagner 1997; Hines and Lütjering 1999).

More specifically, many of these fatigue studies have considered titanium alloys with fully lamellar or duplex (equiaxed primary α + lamellar) microstructures. These studies have shown that crack initiation in the lamellar microstructures occurs mainly across colonies of aligned α platelets (Wells and Sullivan 1969; Eylon and Pierce 1976; Eylon and Hall 1977; Postans and Jeal 1977; Ruppen et al. 1979; Bania et al. 1982; Wojcik et al. 1988; Dowson et al. 1992; Evans and Bache 1994; Lütjering et al. 1996; Wagner 1997; Hines and Lütjering 1999).

The resulting microcracks are faceted with cleavage-like appearances, as are microcracks that initiate in equiaxed primary α (Neal and Blenkinsop 1976). Although these faceted cracks were initially thought to be due to cleavage, it is now more or less accepted that they are caused by intense shear in {0002} slip bands (Wojcik et al. 1988; Evans and Bache 1994; Bache et al. 1998).

Under constant amplitude (CA) fatigue loading the faceted cracks provide no evidence of cyclic crack growth, e.g. fatigue striations. However, fatigue crack growth tests using CA + intermittent spike loading result in progression markings within single facets (Paton et al. 1976; Pilchak et al. 2009). These progression markings demonstrate that the facets developed during many loading cycles.

R. Wanhill and S. Barter, *Fatigue of Beta Processed and Beta Heat-treated Titanium Alloys*, SpringerBriefs in Applied Sciences and Technology, DOI: 10.1007/978-94-007-2524-9_3, © The Author(s) 2012

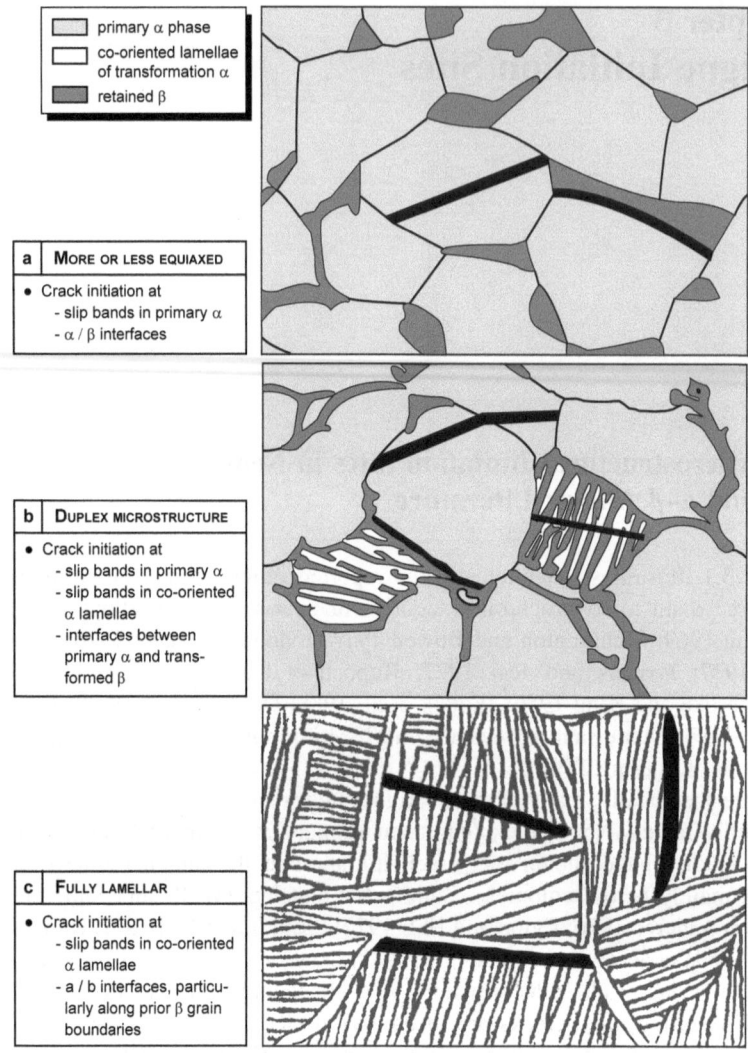

Fig. 3.1 Fatigue crack initiation sites in near-α and α–β titanium alloys

Progression markings can also be detected on fracture surface facets from flight simulation fatigue crack growth tests on the β annealed Ti-6Al-4V ELI thick plate acquired for the DSTO–NLR joint programme, e.g. Fig. 3.2. This is important because the DSTO-EBA fatigue life assessment method requires short-to-long crack growth data using marker loads and Quantitative Fractography (QF), see Table 1.1.

Fig. 3.2 Flight simulation progression markings on a fracture facet from a fatigue crack growth specimen taken from the β annealed Ti-6Al-4V ELI thick plate acquired for the DSTO–NLR joint programme

3.2 Microstructural Initiation Sites: β Annealed Ti-6Al-4V ELI Thick Plate

Some preliminary Low-Cycle Fatigue (LCF) tests were done on cylindrical specimens machined from the β annealed Ti-6Al-4V ELI plate acquired for the DSTO–NLR joint programme. After testing, the surface of one of the specimens was polished and etched to reveal the microstructural locations of surface microcracks. Optical images of the microcracks were obtained using a deep focus image processing technique developed by the DSTO (Goldsmith 2000). Figure 3.3 shows some examples.

Most of the microcracks initiated across colonies of aligned α platelets, as would be expected from previous studies, see Sect. 3.1. Some of these cracks extended with little or no deflection across two or more colonies and their boundaries, and a few also crossed the grain boundary α. There were also cracks along the interfaces between α lamellae and retained β, including colony boundaries; and at least one crack ran along the interface between grain boundary α and colonies of aligned α lamellae (near the top right corner of Fig. 3.3).

Since some microcracks can cross colony boundaries and prior β grain boundaries with little or no deflection, we may infer from the studies by Wojcik et al. (1988), Evans and Bache (1994) and Bache et al. (1998) that these boundaries are not always effective barriers to slip.

Fig. 3.3 Microcracks on the
cylindrical surface of an
LCF-tested specimen taken
from the β annealed Ti-6Al-
4V ELI thick plate acquired
for the DSTO–NLR joint
programme (Kroll's etch)

3.3 Metallurgical Defects

The occurrence of inclusion-type defects in ingot metallurgy titanium alloys is
rare, mainly because the ingots are obtained using high purity materials and
multiple vacuum-arc melting. Titanium powder compacts are, however, suscep-
tible to inclusion defects because of powder contamination by foreign particles
that are not subsequently melted (Kerr et al. 1976).

Costa et al. (1990) reviewed LCF failures due to metallurgical defects in ingot
metallurgy and $(\alpha + \beta)$ processed titanium aeroengine discs. Figure 3.4 classifies
the defect types and categories and their relative frequency of occurrence in 22 in-
service discs:

(1) Type I defects are regions of α phase stabilized by high concentrations of the
    interstitial elements nitrogen and oxygen (sometimes called HIDs = High
    Interstitial Defects). Category 1 defects are very hard and brittle; category 2
    defects have lower hardness, but still higher than that of the matrix. The
    sources of type I defects are high melting point particles of titanium nitride,
    titanium oxide or complex oxynitride coming from titanium sponge ("burnt"
    sponge), master alloy additions or revert (recycled) material.
(2) Type II defects are regions containing an excessive amount of primary α that is
    abnormally stabilized by segregation of metallic elements, notably aluminium
    or titanium. Category 3 defects have hardnesses only slightly above that of the
    matrix; category 4 defects can have very low hardness (these are sometimes
    called LADs = Low Alloy Defects).

While important, this information has to be put into perspective. Titanium disc
failures are rare: of the six engine manufacturers visited by Costa's Review Team,

	Type I Defects		Type II Defects	
	Category 1	Category 2	Category 3	Category 4
Metallurgical observations (typical)	• Nitrogen stabilized hard alpha zone (zone 2) encasing large spongy-appearing void (zone 1) • Alpha case surrounded by enlarged or blocky alpha grains or platelets (zone 3) 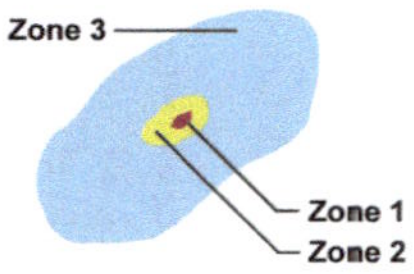	• Small or no voids (zone 1) • No hard alpha zone • Large area of nitrogen stabilized enlarged and elongated alpha grains or platelets (zone 3) 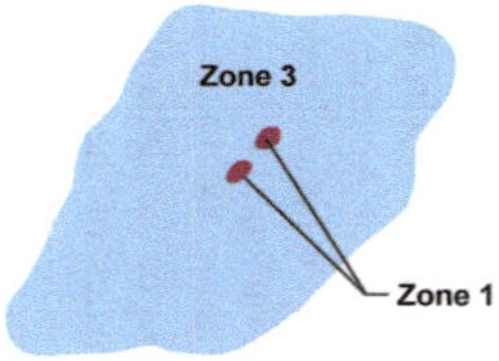	• Microvoids (zone 1) • Low or no elevated nitrogen or oxygen concentration • Large area of aluminium stabilized enlarged and elongated alpha grains or platelets (zone 3) 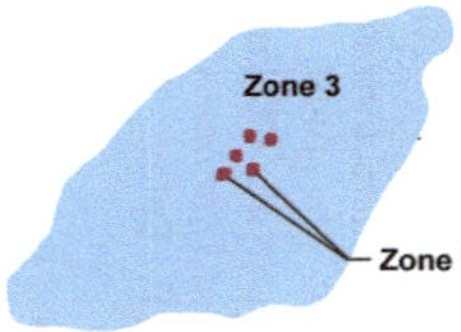	• Pure elemental segregation of Ti or Al (zone3) 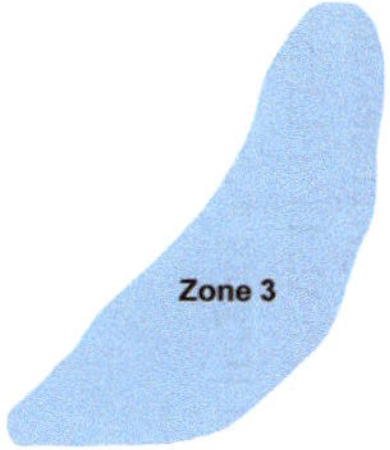
Zone hardness*	• Zone 2 = RC 65-80 • Zone 3 = RC 55-70	RC 45-65	RC 35-45	RC 12
Zone shape	All Zones Ellipsoid Shaped as Per the Direction of Work			
Most probable cause	Burnt titanium sponge (source material for ingot production)	Contaminated weldment or contaminated revert material entering ingot	Inclusion drop-in during ingot production	Improperly melted/homogenized alloy or a solidification pipe
Defects in 22 in-service discs	41%	41%	14%	4%

Increasing difficulty to detect by ultrasonic testing

* RC = Rockwell hardness "C" scale

Fig. 3.4   Classification of metallurgical defects causing fatigue failures in titanium alloy aeroengine discs (Costa et al. 1990)

four had a combined total of 25 discs that cracked or failed in service owing to metallurgical defects. This is a very small number set against the thousands of discs in service up to the time of the review (1990).

Furthermore, as stated at the beginning of this section, ingot metallurgy alloys are produced using high purity materials and multiple vacuum-arc melting. Hence it is most unlikely that fatigue-initiating metallurgical defects will be present in a premium quality material like β annealed Ti-6Al-4V ELI plate, although defects such as machining tears, weld defects and forging laps are possible in manufactured components.

References

M.R. Bache, W.J. Evans, V. Randle, R.J. Wilson, Characterization of mechanical anisotropy in titanium alloys. Mater. Sci. Eng. A **A257**, 139–144 (1998)

P.J. Bania, L.R. Bidwell, J.A. Hall, D. Eylon, A.K. Chakrabarti, Fracture—microstructure relationships in titanium alloys, in *Titanium and Titanium Alloys, Scientific and Technological Aspects*, vol. 1, ed. by J.C. Williams, A.F. Belov (Plenum Press, New York, 1982), pp. 663–677

R.K. Bolingbroke, J.E. King, The growth of short fatigue cracks in titanium alloys IMI550 and IMI318, in *Small Fatigue Cracks*, ed. by R.O. Ritchie, J. Lankford (The Metallurgical Society, Inc., Warrendale, 1986), pp. 129–144

J.G. Costa, R.E. Gonzalez, R.E. Guyotte, D.P. Salvano, T. Swift, R.J. Koenig, Titanium Rotating Components Review Team Report, United States of America Federal Aviation Administration, Aircraft Certification Service, Engine and Propeller Directorate (1990)

X. Demulsant, J. Mendez, Microstructural effects on small fatigue crack initiation and growth in Ti6Al4V alloys. Fatigue Fract. Eng. Mater. Struct. **18**, 1483–1497 (1995)

A.L. Dowson, A.C. Hollis, C.J. Beevers, The effect of the alpha-phase volume fraction and stress ratio on the fatigue crack growth characteristics of the near-alpha IMI 834 Ti alloy. Int. J. Fatigue **14**, 262–270 (1992)

W.J. Evans, M.R. Bache, Dwell-sensitive fatigue under biaxial loads in the near-alpha titanium alloy IMI685. Int. J. Fatigue **16**, 443–452 (1994)

D. Eylon, J.A. Hall, Fatigue behavior of beta processed titanium alloy IMI 685. Metall. Trans. A **8A**, 981–990 (1977)

D. Eylon, C.M. Pierce, Effect of microstructure on notch fatigue properties of Ti-6Al-4V. Metall. Trans. A **7A**, 111–121 (1976)

N.T. Goldsmith, Deep focus; a digital image processing technique to produce improved focal depth in light microscopy. Image Anal. Stereol. **19**, 163–167 (2000)

J.A. Hines, G. Lütjering, Propagation of microcracks at stress amplitudes below the conventional fatigue limit in Ti-6Al-4V. Fatigue Fract. Eng. Mater. Struct. **22**, 657–665 (1999)

W.R. Kerr, D. Eylon, J.A. Hall, On the correlation of specific fracture surface and metallographic features by precision sectioning in titanium alloys. Metall. Trans. A **7A**, 1477–1480 (1976)

G. Lütjering, A. Gysler, J. Albrecht, Influence of microstructure on fatigue resistance, in *Fatigue '96*, vol. II, ed. by G. Lütjering, H. Nowack (Elsevier Science Ltd, Oxford, 1996), pp. 893–904

D.F. Neal, P.A. Blenkinsop, Internal fatigue origins in α–β titanium alloys. Acta Metall. **24**, 59–63 (1976)

N.E. Paton, J.C. Williams, J.C. Chesnutt, A.W. Thompson, The effects of microstructure on the fatigue and fracture of commercial titanium alloys, AGARD Conference Proceedings No. 185,

Advisory Group for Aerospace Research and Development, Neuilly-sur-Seine, 1976, pp. 4-1–4-14

A.L. Pilchak, A. Bhattacharjee, A.H. Rosenberger, J.C. Williams, Low ΔK faceted crack growth in titanium alloys. Int. J. Fatigue **31**, 989–994 (2009)

P.J. Postans, R.H. Jeal, Dependence of crack growth performance upon structure in β processed titanium alloys, in *Forging and Properties of Aerospace Materials* (The Metals Society, London, 1977), pp. 192–198

J. Ruppen, P. Bhowal, D. Eylon, A.J. McEvily, On the process of subsurface fatigue crack initiation in Ti-6Al-4V, in *Fatigue Mechanisms, ASTM STP 675*, ed. by J. Fong (American Society for Testing and Materials, Philadelphia, 1979), pp. 47–68

C.A. Stubbington, A.W. Bowen, Improvements in the fatigue strength of Ti-6Al-4V through microstructure control. J. Mater. Sci. **9**, 941–947 (1974)

L. Wagner, Fatigue life behavior, in *ASM Handbook Volume 19 Fatigue and Fracture, Second Printing*, ed. by S.R. Lampman et al. (ASM International, Materials Park, 1997), pp. 837–845

C.H. Wells, C.P. Sullivan, Low-cycle fatigue crack initiation in Ti-6Al-4V. Trans. ASM **62**, 263–270 (1969)

C.C. Wojcik, K.S. Chan, D.A. Koss, Stage I fatigue crack propagation in a titanium alloy. Acta Metall. **36**, 1261–1270 (1988)

Chapter 4
Fatigue Initiation Lives

4.1 Microstructural Factors

The fatigue life behaviour of titanium alloys depends on several microstructural factors whose importance differs for conventionally ($\alpha + \beta$) processed and heat-treated alloys and β processed and β heat-treated alloys. For ($\alpha + \beta$) processed and heat-treated alloys the significant microstructural factors are:

(1) *Primary α grain size.* A smaller primary α grain size increases the High-Cycle Fatigue (HCF) strength (Turner and Roberts 1968; Lucas and Konieczny 1971; Lucas 1973; Bowen and Stubbington 1973; Stubbington and Bowen 1974; Lütjering et al. 1996; Wagner 1997).

This correlation has been explained as follows. Firstly, a smaller primary α grain size results in a higher yield strength, such that higher stresses are required to initiate slip in the primary α and cause slip band fatigue cracks (Lütjering et al. 1996; Wagner 1997). Secondly, any cracks that do form will be shorter and easier to arrest at the microstructural barriers provided by α/α grain boundaries (Demulsant and Mendez 1995) and $\alpha/(\alpha + \beta)$ grain boundaries in duplex microstructures (Bolingbroke and King 1986; Demulsant and Mendez 1995).

(2) *Material texture.* Crystallographic alignments of primary α grains can have large effects on HCF strength (Stubbington and Bowen 1972; Bowen and Stubbington 1973; Larson and Zarkades 1976; Peters et al. 1984; Lütjering and Wagner 1988).

In general, the highest fatigue strength is obtained for strong textures when the {0002} planes are parallel to the principal loading direction (Peters et al. 1984). This orientation inhibits slip band cracking in the primary α (Lütjering and Wagner 1988).

R. Wanhill and S. Barter, *Fatigue of Beta Processed and Beta Heat-treated Titanium Alloys*, SpringerBriefs in Applied Sciences and Technology, DOI: 10.1007/978-94-007-2524-9_4, © The Author(s) 2012

(3) *Oxygen content and primary α hardness.* Higher oxygen contents increase the yield strength and hardness of primary α (Beevers and Robinson 1969; Curtis et al. 1969; Sargent and Conrad 1972; Williams et al. 1972; Robinson and Beevers 1973; Yoder et al. 1984). This correlates with an increase in HCF strength provided that crack initiation occurs in the primary α, and not—as can occur—in aligned α lamellae in duplex microstructures (Lütjering and Wagner 1988; Lütjering et al. 1996; Wagner 1997). See Sect. 4.2.1 also.

The effect of oxygen on primary α fatigue strength is threefold. Firstly, a higher yield strength means that higher stresses are required to cause slip band fatigue cracks: see (1) above. Secondly, oxygen restricts $\{0002\}$ slip, compared to $\{10\bar{1}0\}$ and $\{10\bar{1}1\}$ slip, up to oxygen contents ~ 0.5 wt.% (Curtis et al. 1969; Sargent and Conrad 1972; Williams et al. 1972); and since fatigue crack initiation in primary α is largely due to intense slip on $\{0002\}$, increased oxygen content would be expected to inhibit cracking. Thirdly, increasing oxygen content changes the slip character from wavy to planar (Curtis et al. 1969; Williams et al. 1972; Kahveci and Welsch 1989). This means that cross-slip becomes more difficult, and it is well known that cross-slip promotes slip band fatigue cracking (McEvily and Johnston 1966).

For β processed and β heat-treated alloys the significant microstructural factors are:

(4) *Colony and lamella sizes.* A smaller colony size increases the LCF (Eylon and Hall 1977) and HCF (Farthing 1989) strengths, and narrower aligned α lamellae within the colonies especially increase the HCF fatigue strength (Wagner 1997).

(5) *Prior β grain size.* A smaller prior β grain size increases the LCF and HCF strengths (Wagner 1997; Evans 1999).

The trends in (4) and (5) may be attributed firstly to less easy nucleation of slip band fatigue cracks through the aligned α lamellae, since the cracks have to cross through more of the retained β "ribs". Secondly, any cracks that do form will sooner reach the microstructural barriers formed by colony boundaries and prior β grain boundaries, i.e. these cracks will be shorter and easier to arrest.

In view of all the foregoing microstructural factors, it is not easy to compare the fatigue strengths of conventionally $(\alpha + \beta)$ processed and heat-treated alloys and β processed and β heat-treated alloys. However, some trends have been observed. These are discussed in Sect. 4.2.

4.2 Fatigue Life–Microstructure Trends

In this context the available comparisons of fatigue life data are for conventionally $(\alpha + \beta)$ processed alloys subsequently $(\alpha + \beta)$ or β heat-treated. In each case the cracks grew from surfaces free of significant mechanical defects.

4.2.1 Unnotched (Smooth Specimen) Fatigue

Figures 4.1, 4.2 and 4.3 compare unnotched LCF data for Ti-6Al-4V in several microstructural conditions, including the as-received mill annealed condition. The following trends may be observed:

(1) A coarse prior β grain size (and hence larger colonies and aligned α platelets) consistently results in the lowest fatigue curve, see Figs. 4.1 and 4.2.
(2) A fine prior β grain size (and hence smaller colonies and aligned α platelets) results in a fatigue curve

- slightly lower than that of the original mill annealed microstructure, Fig. 4.1,
- significantly lower than that of duplex microstructures containing 10–30% primary α, but slightly better than that of a duplex microstructure containing 82% primary α, Fig. 4.3.

Figures 4.4 and 4.5 show additional data for fully lamellar and duplex microstructures. These data extend the fatigue lives into the HCF regime and show fatigue curve crossovers between 10^4 and 10^5 cycles.

These LCF→HCF trends can be explained using a rationale by Hines and Lütjering (1999). They distinguished between the LCF and HCF regimes as follows.

Under LCF conditions the key feature is the microstructural scale. Fully lamellar microstructures have much larger scales than mill annealed or duplex microstructures, and so the slip band length is much longer. This feature causes earlier fatigue crack initiation and therefore lower fatigue curves. (Note that this would be especially true for a coarse prior β grain size.)

Under HCF conditions the maximum stresses are relatively low, and the slip band length becomes less important than the intrinsic lattice resistance to dislocation motion. For duplex microstructures the lattice resistance to dislocation motion is affected by alloying element partitioning: elements such as aluminium and oxygen tend to partition into the primary α during heat treatment, thereby weakening the β matrix and the subsequent lamellar microstructure. In this way the transformed β in a duplex microstructure can be weaker than that in a fully lamellar microstructure, leading to earlier fatigue crack initiation and LCF→HCF fatigue curve crossovers.

It is also worth noting that Lütjering et al. (1996) and Wagner (1997) found that larger amounts of primary α in duplex microstructures resulted in lower LCF→HCF curves, e.g. Fig. 4.5. However, the LCF results in Fig. 4.3 are not entirely consistent with this.

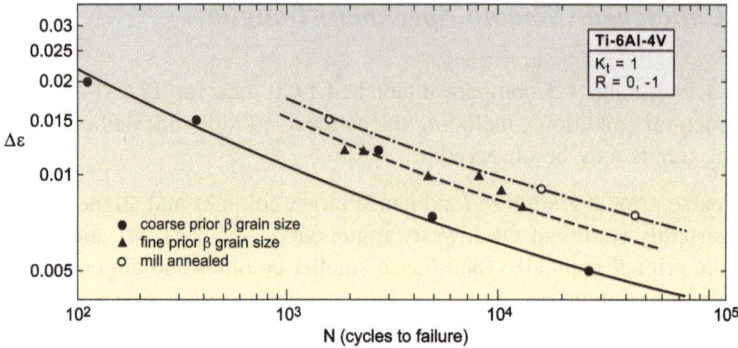

Fig. 4.1 ε–N fatigue curves for Ti-6Al-4V in three microstructural conditions: fully lamellar (coarse prior β grain size), fully lamellar (fine prior β grain size) and mill annealed (Evans 1999)

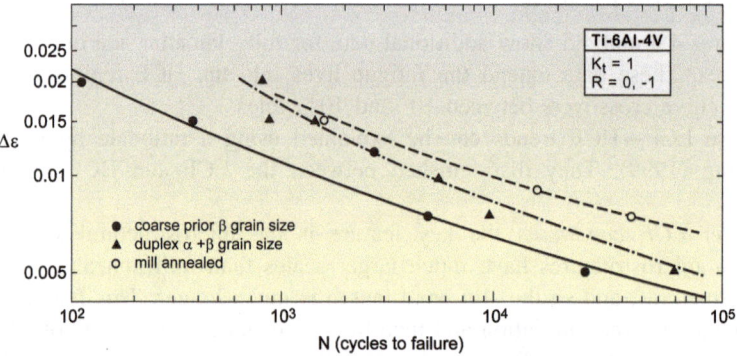

Fig. 4.2 ε–N fatigue curves for Ti-6Al-4 V in three microstructural conditions: fully lamellar (coarse prior β grain size), duplex (82% α) and mill annealed (Evans 1999)

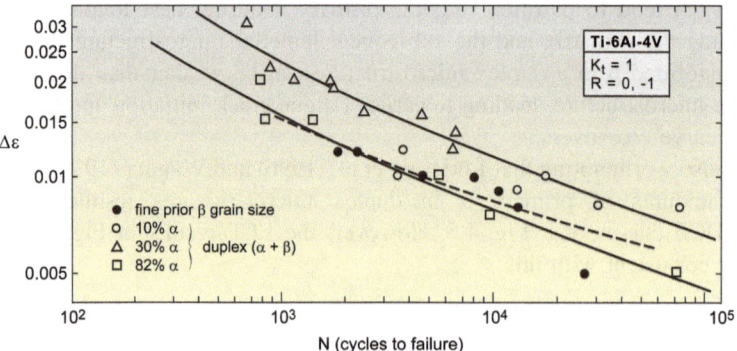

Fig. 4.3 ε–N fatigue curves for Ti-6Al-4V in four microstructural conditions: fully lamellar (fine prior β grain size) and duplex (10% α, 30% α, 82% α) (Evans 1999)

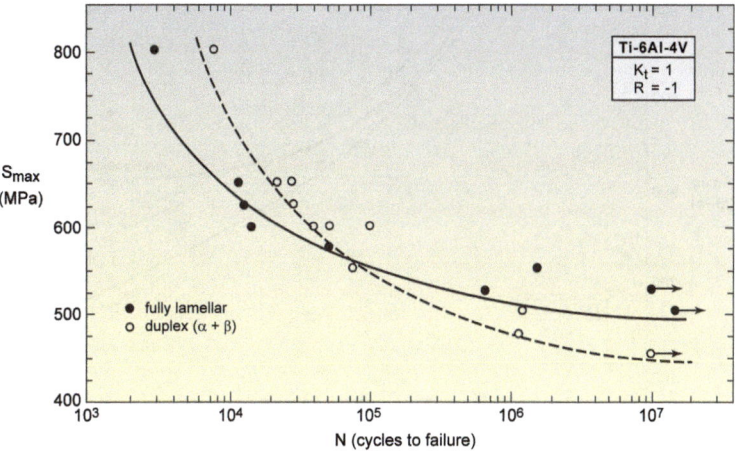

Fig. 4.4 S–N fatigue curves for Ti-6Al-4V in two microstructural conditions: fully lamellar and duplex (35% α) (Hines and Lütjering 1999)

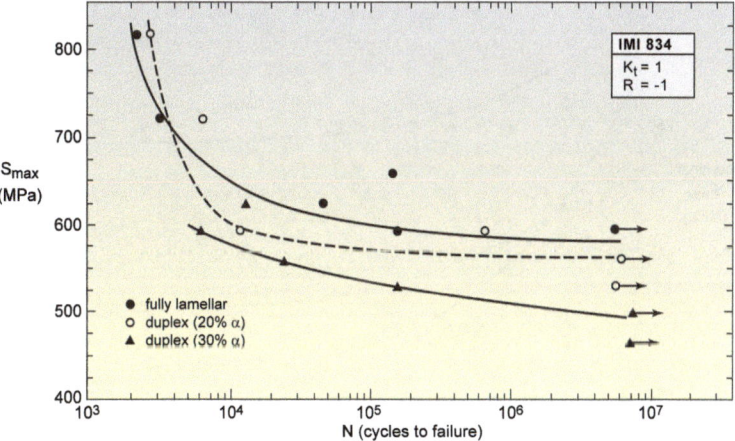

Fig. 4.5 S–N fatigue curves for IMI 834 (Ti-5.8Al-4.0Sn-3.5Zr-0.7Nb-0.5Mo-0.35Si-0.06C) in three microstructural conditions: fully lamellar and duplex (20% α, 30% α) (Lütjering et al. 1999)

4.2.2 Notched Fatigue

Figures 4.6 and 4.7 are from two investigations comparing notched LCF→HCF curves for Ti-6Al-4V in several microstructural conditions, including the as-received mill annealed conditions (Eylon and Pierce 1976; Evans 1999). These results are very interesting, not least because the trends in each figure are different:

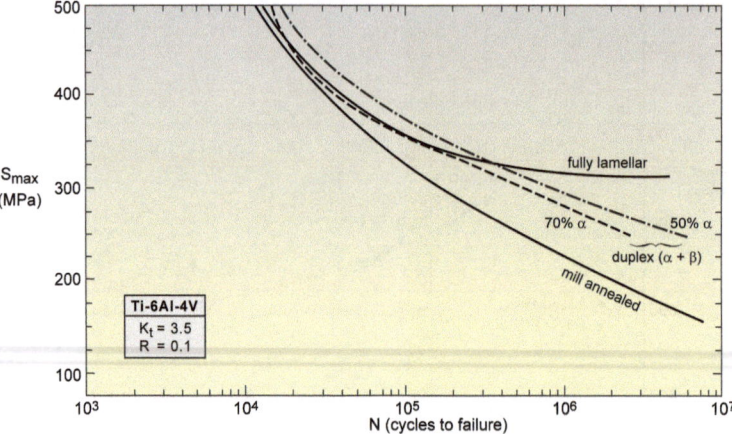

Fig. 4.6 S–N notched fatigue curves for Ti-6Al-4V plate in four microstructural conditions: fully lamellar, duplex (50% α, 70% α) and mill annealed (Eylon and Pierce 1976)

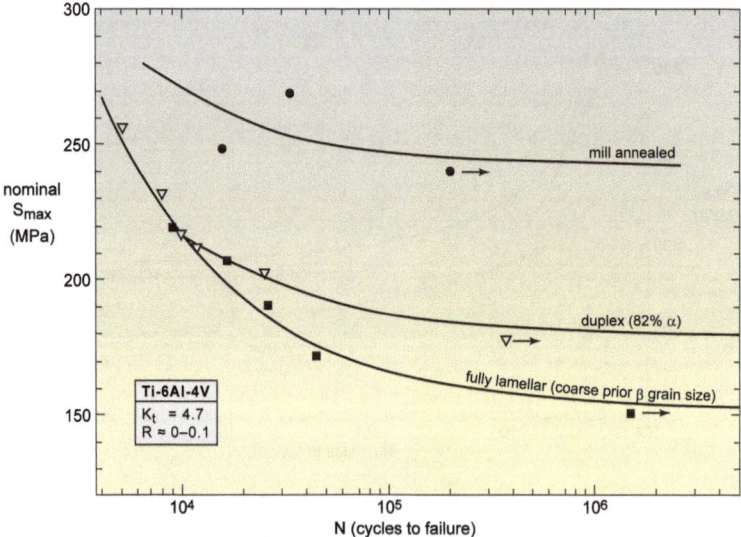

Fig. 4.7 S–N notched fatigue curves for Ti-6Al-4V plate in three microstructural conditions: fully lamellar, duplex (82% α) and mill annealed (Evans 1999)

(1) Figure 4.6 shows that the notched HCF strength of a fully lamellar micro-structure was much better than the notched HCF strengths of duplex or mill annealed microstructures.

(2) Figure 4.7 indicates that the notched HCF strength of a mill annealed microstructure was far better than that of the other microstructures.

The explanation for these differences most probably lies in the microstructural scale. The rankings in Fig. 4.6 are probably due to the fine prior β grain size, about 0.25 mm, for the fully lamellar microstructure, and large elongated primary α grains, 0.1–0.15 mm, in the mill annealed microstructure. On the other hand, the rankings in Fig. 4.7 are probably due to the coarse prior β grain size, about 1.5 mm, for the fully lamellar microstructure, and much smaller primary α grains, 0.02–0.04 mm, in the mill annealed microstructure.

We note that this explanation is apparently at variance with Hines' and Lütjering's rationale for unnotched fatigue rankings, discussed in Sect. 4.2.1, since they attributed only LCF rankings to the microstructural scale. However, the coarse mill annealed microstructure of the material tested by Eylon and Pierce (1976) may well be unrepresentative with respect to later mill products.

4.3 Summary

As stated earlier, it is not easy to compare the fatigue strengths of conventionally $(\alpha + \beta)$ processed and heat-treated alloys and β processed and β heat-treated alloys. However, from the information in Sects. 4.2.1 and 4.2.2 it appears likely that the β annealed Ti-6Al-4V ELI thick plate, with its rather coarse prior β grain size (averaging about 1.2 mm) will have relatively low LCF and HCF strengths. The relatively low oxygen content of the ELI material (0.13 max. wt.% compared to 0.2 max. wt.% for normal grade Ti-6Al-4V) could also be detrimental to the HCF strength, see Starke and Lütjering (1979).

References

C.J. Beevers, J.L. Robinson, Some observations on the influence of oxygen content on the fatigue behaviour of α-titanium. J. Less-Common Met. **17**, 345–352 (1969)

A.W. Bowen, C.A. Stubbington, The effect of $\alpha + \beta$ working on the fatigue and tensile properties of Ti-6Al-4V bars, in *Titanium Science and Technology*, ed. by R.I. Jaffee, H.M. Burte (Plenum Press, New York, 1973), pp. 2097–2108

R.K. Bolingbroke, J.E. King, The growth of short fatigue cracks in titanium alloys IMI550 and IMI318, in *Small Fatigue Cracks*, ed. by R.O. Ritchie, J. Lankford (The Metallurgical Society, Inc., Warrendale, 1986), pp. 129–144

R.E. Curtis, R.R. Boyer, J.C. Williams, Relationship between composition, microstructure, and stress corrosion cracking (in salt solution) in titanium alloys. Trans. ASM **62**, 457–469 (1969)

X. Demulsant, J. Mendez, Microstructural effects on small fatigue crack initiation and growth in Ti6Al4V alloys. Fatigue Fract. Eng. Mater. Struct. **18**, 1483–1497 (1995)

W.J. Evans, Microstructure and the development of fatigue cracks at notches. Mater. Sci. Eng. A **A263**, 160–175 (1999)

D. Eylon, J.A. Hall, Fatigue behavior of beta processed titanium alloy IMI 685. Metall. Trans. A **8A**, 981–990 (1977)

D. Eylon, C.M. Pierce, Effect of microstructure on notch fatigue properties of Ti-6Al-4V. Metall. Trans. A **7A**, 111–121 (1976)

T.W. Farthing, Titanium—the producer's view, in *Metals Fight Back Conference—Advanced Metallic Alloys for Aerospace Applications*, Shephard Conferences and Exhibitions, Slough (1989)

J.A. Hines, G. Lütjering, Propagation of microcracks at stress amplitudes below the conventional fatigue limit in Ti-6Al-4V. Fatigue Fract. Eng. Mater. Struct. **22**, 657–665 (1999)

A.I. Kahveci, G. Welsch, Effects of oxygen on phase composition and strength of Ti-6Al-4V Alloy, in *Proceedings of the Sixth World Conference on Titanium*, vol. 1, ed. by P. Lacombé, R. Tricot, G. Béranger (Les Éditions de Physique, Paris, 1989), pp. 339–343

F.R. Larson, A. Zarkades, Improved fatigue life in titanium through texture control, in *Texture and the Properties of Materials*, ed. by G.J. Davies, I.L. Dillamore, R.C. Hudd, J.S. Kallend (The Metals Society, London 1976), pp. 210–216

J.J. Lucas, P.P. Konieczny, Relationship between alpha grain size and crack initiation fatigue strength in Ti-6Al-4V. Metall. Trans. **2**, 911–912 (1971)

J.J. Lucas, Improvements in the fatigue strength of Ti-6Al-4V forgings, in *Titanium Science and Technology*, ed. by R.I. Jaffee, H.M. Burte (Plenum Press, New York, 1973), pp. 2081–2095

G. Lütjering, A. Gysler, J. Albrecht, Influence of microstructure on fatigue resistance, in *Fatigue '96*, vol. II, ed. by G. Lütjering, H. Nowack (Elsevier Science Ltd, Oxford, 1996), pp. 893–904

G. Lütjering, L. Wagner, Influence of texture on fatigue properties of titanium alloys, in *Directional Properties of Materials*, ed. by H.J. Bunge (DGM Informationsgesellschaft mbH, Oberursel, 1988), pp. 177–188

A.J. McEvily Jr., T.L. Johnston, The role of cross-slip in brittle fracture and fatigue, in *Proceedings of the First International Conference on Fracture*, vol. 2, ed. by T. Yokobori, T. Kawasaki, J.L. Swedlow (The Japanese Society for Strength and Fracture of Materials, Japan, 1966), pp. 515–546

M. Peters, A. Gysler, G. Lütjering, Influence of texture on fatigue properties of Ti-6Al-4V. Metall. Trans. A **15A**, 1597–1605 (1984)

J.L. Robinson, C.J. Beevers, The effects of load ratio, interstitial content, and grain size on low-stress fatigue-crack propagation in α-titanium. Met. Sci. J. **7**, 153–159 (1973)

G.A. Sargent, H. Conrad, On the strengthening of titanium by oxygen. Scripta METALLURGICA **6**, 1099–1101 (1972)

E.A. Starke Jr., G. Lütjering, Cyclic plastic deformation and microstructure, in *Fatigue and Microstructure* (American Society for Metals, Metals Park, 1979), pp. 205–243

C.A. Stubbington, A.W. Bowen, The effect of section size on the fatigue properties of Ti-6Al-4V bars, Royal Aircraft Establishment Technical Report TR72091, Procurement Executive, Ministry of Defence, Farnborough, UK (1972)

C.A. Stubbington, A.W. Bowen, Improvements in the fatigue strength of Ti-6Al-4V through microstructure control. J. Mater. Sci. **9**, 941–947 (1974)

N.G. Turner, W.T. Roberts, Fatigue behavior of titanium. Trans. Metall. Soc. AIME **242**, 1223–1230 (1968)

J.C. Williams, A.W. Sommer, P.P. Tung, The influence of oxygen concentration on the internal stress and dislocation arrangements in α titanium. Metall. Trans. **3**, 2979–2984 (1972)

L. Wagner, Fatigue life behavior, in *ASM Handbook Volume 19 Fatigue and Fracture, Second Printing*, ed. by S.R. Lampman et al. (ASM International, Materials Park, 1997), pp. 837–845

G.R. Yoder, F.H. Froes, D. Eylon, Effect of microstructure, strength, and oxygen content on fatigue crack growth rate of Ti-4.5Al-5.0Mo-1.5Cr (CORONA 5). Metall. Trans. A **15A**, 183–197 (1984)

Chapter 5
Short/Small Fatigue Crack Growth

5.1 Introduction

As is well known, there is considerable evidence that short/small fatigue cracks in metals grow at faster rates and lower nominal ΔK values than those characteristic of long/large cracks (Ritchie and Suresh 1983; Suresh and Ritchie 1984). In particular, short/small cracks can grow at ΔK values well below the long/large crack growth threshold, ΔK_{th}.

Short/small fatigue crack growth is a complex subject, owing to the variety of factors that can affect the crack behaviour (McClung et al. 1996). For many aerospace alloys the differences in crack growth behaviour between short/small cracks and long/large cracks disappear for crack sizes larger than 0.25–0.5 mm (Anstee 1983; Anstee and Edwards 1983). However, there are exceptions, including β processed and β heat-treated titanium alloys, as will be discussed in Sect. 5.2.

5.1.1 Significance of Short/Small Cracks

The significance of short/small fatigue cracks during the total life of a component or structure depends on whether the design and service lives are intended to be in the LCF or HCF regimes. To begin with, we may consider the fatigue life to consist of three stages:

$$N_t = N_i + N_{sc} + N_{lc} \qquad (5.1)$$

where N_t is the total life; N_i is the life to crack initiation (sometimes contested as non-existent); N_{sc} is the life spent in growing a short/small crack; and N_{lc} is the life spent in growing a long (large) crack.

Whether or not N_i exists, most of the life is spent before entering the long (large) crack growth stage. For LCF the estimates vary from 70–90% of N_t, and for

R. Wanhill and S. Barter, *Fatigue of Beta Processed and Beta Heat-treated Titanium Alloys*, SpringerBriefs in Applied Sciences and Technology, DOI: 10.1007/978-94-007-2524-9_5, © The Author(s) 2012

HCF it can exceed 95% of N_t, e.g. Schijve (1967), Bania et al. (1982), James and Knott (1985) and Demulsant and Mendez (1995).

From these estimates we may reasonably conclude that short/small crack growth should be investigated as part of a study of fatigue life assessment methods.

5.1.2 Definitions of Short and Small Cracks

The terms "short crack" and "small crack" both appear in the literature, sometimes virtually as synonyms. However, these terms have acquired distinct meanings, particularly in the United States (McClung et al. 1996). A crack defined as "short" need have only one physical dimension that is small, but a "small" crack's dimensions are all small.

There are also different types of short or small cracks (Ritchie and Suresh 1983; McClung et al. 1996):

(1) *Microstructurally short or small cracks.* These are cracks with one or more dimensions smaller than a characteristic microstructural dimension, usually based on the grain size. For titanium alloys this could be the primary α grain size in conventionally $(\alpha + \beta)$ processed and heat-treated alloys; and the prior β grain size or colony size in β processed and β heat-treated alloys.
(2) *Mechanically short or small cracks.* These are cracks with one or more dimensions smaller than characteristic mechanical dimensions. These dimensions typically define regions of plastic deformation, e.g. crack tip plastic zones or local plasticity at the roots of notches or other stress concentrations.
(3) *Physically/chemically short or small cracks.* These are cracks with one or more dimensions larger than characteristic microstructural and/or mechanical dimensions that nevertheless can grow significantly faster than truly large cracks at comparable ΔK values.

There is general agreement about the nomenclature for types (1) and (2). Whether type (3) should be called *physically* or *chemically* short or small is not entirely clear, since justification for the existence of this third type depends largely on corrosion fatigue results, see Ritchie and Suresh (1983) and McClung et al. (1996).

5.1.3 Size Criteria

Table 5.1 gives suggestions for classifying small and large crack sizes according to microstructural and mechanical criteria (Taylor 1986; McClung et al. 1996):

(1) Cracks are generally considered to be microstructurally small when (a) their size is less than 5–10 times the microstructural unit size, M, or (b) the crack tip plastic zone size is less than or equal to M. For titanium alloys M may be the primary α or prior β grain size, or the colony size, as noted in Sect. 5.1.1.

Table 5.1 Size criteria for small cracks (McClung et al. 1996)

Microstructural size	Mechanical size	
	Large: $a/r_p > 4$–20 (SSY)	Small: $a/r_p < 4$–20 (ISY and LSY)
Large: $a/M > 5$–10 and $r_p/M \gg 1$	Mechanically and microstructurally large (LEFM valid)	Mechanically small but microstructurally large (may need EPFM)
Small: $a/M < 5$–10 and $r_p/M \sim 1$	Mechanically large but microstructurally small	Mechanically and microstructurally small

a crack size; r_p crack tip plastic zone size; *M* microstructural unit size; *SSY* small scale yielding; *ISY* intermediate scale yielding; *LSY* large scale yielding; *LEFM* linear elastic fracture mechanics; *EPFM* elastic–plastic fracture mechanics

(2) Cracks often behave in a mechanically small manner when the ratio of crack size to crack tip plastic zone size is less than 4–20.

Another approach to describing fatigue behaviour in terms of crack size is illustrated in Fig. 5.1. This shows three regimes of fatigue behaviour: crack initiation; irregular crack growth and coalescence of any neighbouring short or small cracks; and more or less regular growth of short-to-long (or small-to-large) cracks. These three regimes are defined by both crack size and cyclic stress levels:

- Firstly, relatively high stress amplitudes that exceed the fatigue stress range endurance limit, ΔS_e, are generally necessary for crack initiation,[1] and one or several microcracks may be initiated. These cracks must overcome microstructural barriers in order to grow, and are therefore classified as microstructurally short or small until they reach a certain size, a_1, defined by the criteria in Table 5.1 and (1) above.
- Between a_1 and a_2, defined by the criteria in Table 5.1 and (2) above, the cracks are classified as mechanically short or small. This means that their behaviour cannot be properly described by LEFM. As Fig. 5.1 indicates, these cracks first grow irregularly, i.e. intermittently along the crack front, including the coalescence of any neighbouring cracks (Demulsant and Mendez 1995). The irregular crack growth is mainly due to a persistent but lessening influence of microstructural barriers. Also it is no longer necessary that the cyclic stress amplitudes exceed ΔS_e in order to cause further cracking. This phenomenon was first reported by Kitagawa and Takahashi (1976).
- Irregular crack growth gradually gives way to nominally regular crack growth, which depends both on crack size and the cyclic stress levels: higher cyclic stress amplitudes result in transitions to regular crack growth at shorter or smaller crack sizes.
- Beyond a_2 the cracks are classified as long or large. At cyclic stress amplitudes below two-thirds of the cyclic yield stress, σ_y^c, the regular crack growth becomes amenable to description by LEFM.

---

[1]   Crack initiation below ΔS_e occurs in some materials, e.g. carbon steels (De los Rios et al. 1985).

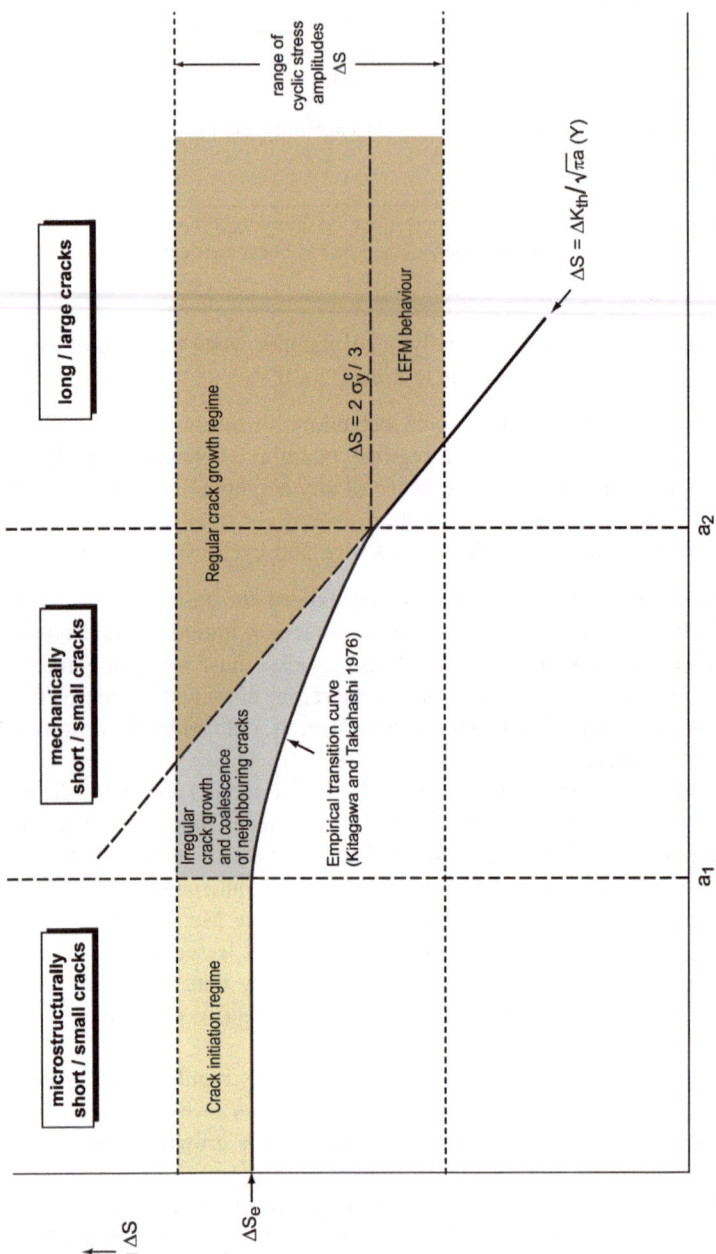

Fig. 5.1 Dependence of fatigue crack behaviour (initiation, irregular growth and coalescence, sustained growth) on cyclic stress amplitudes, ΔS, and crack sizes, a. ΔS_e fatigue stress range endurance limit; ΔK_{th} fatigue crack growth threshold for long/large cracks; Y geometric factor dependent on crack shape and size and specimen or component shape and size; σ_y^c = cyclic yield stress

5.2 Short/Small Fatigue Crack Growth Data from the Literature

Most of the short/small fatigue crack growth data for β processed and β heat-treated titanium alloys have been obtained for the alloy IMI 685, with some data also for IMI 829, IMI 834, Ti65S and IMI 318 (Ti-6Al-4V). There are two reasons for this selection of alloys:

(1) *Alloy composition.* Table 5.2 gives the nominal chemical compositions of the alloys. IMI 685, IMI 829, IMI 834 and Ti65S are near-α alloys, containing predominantly α-stabilizers (aluminium, zirconium, and tin). They are generally β processed and/or β heat-treated, resulting in fully or partially lamellar microstructures of varying coarseness, depending on the processing conditions. On the other hand, IMI 318 (Ti-6Al-4V) is an $\alpha-\beta$ alloy, containing the α-stabilizer aluminium and a substantial amount of the β-stabilizer vanadium. $\alpha-\beta$ alloys have conventionally been ($\alpha + \beta$) processed and heat-treated, as mentioned earlier, in Sect. 5.2.3, because it is more difficult to obtain satisfactory mechanical properties from β processing and/or β heat-treating (Coyne 1970; Green and Minton 1970; Donachie 1982, 2000; Terlinde et al. 2003; Wagner 1997; Evans 1999).

(2) *Precedence.* IMI 685 was the first near-α alloy to be used in the fully lamellar condition, and so its fatigue crack growth properties were of pioneering importance.

Table 5.2 Nominal chemical compositions of several titanium alloys (wt.%)

Near-α alloys	IMI 685	Ti-6Al-5Zr-0.5Mo-0.25Si-0.1Fe
	IMI 829	Ti-5.5Al-3Zr-3.5Sn-1Nb
	IMI 834	Ti-5.8Al-3.5Zr-4Sn-0.7Nb-0.5Mo-0.35Si-0.06C
	Ti 65S	Ti-6Al-5Zr-0.5Mo-0.25Si-0.2Fe-100ppmHf
$\alpha-\beta$ alloy	IMI 318	Ti-6Al-4V-0.15Fe-0.17O

5.2.1 Coarse-Grained Fully Lamellar IMI 685

Figures 5.2 and 5.3 compare microstructurally short fatigue crack growth data with a long crack growth curve for coarse-grained IMI 685 (Hicks and Brown 1982; Brown and Hicks 1983; Hicks et al. 1983). The cracks were up to 3.5 mm long, while the prior β grain size and colony size were about 5 and 1 mm respectively. Although the short crack data encompasses wide ranges in crack growth rates, there are clear trends:

• Crystallographic crack growth across aligned α platelets tended to be faster than non-crystallographic crack growth and cracking along colony boundaries and α/β interfaces.

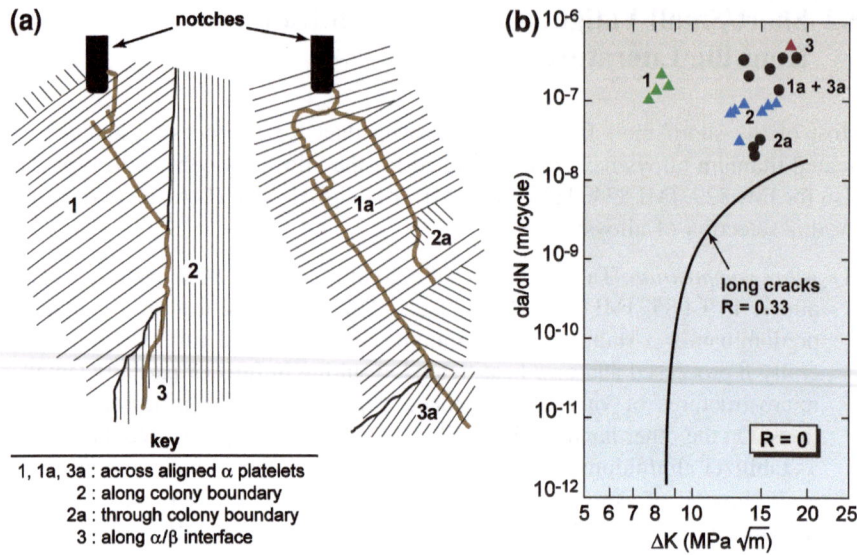

Fig. 5.2 Microstructurally short fatigue crack growth in IMI 685: **a** sketches of actual crack paths; **b** crack growth rates compared to a long crack growth rate curve (Hicks and Brown 1982)

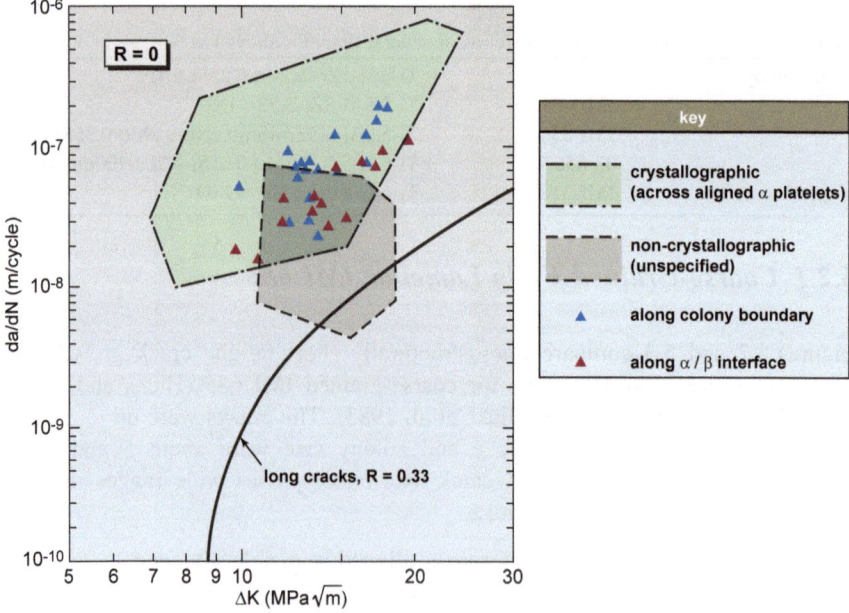

Fig. 5.3 Microstructurally short fatigue crack growth in IMI 685. After Brown and Hicks (1983) and Hicks et al. (1983)

- Short crack growth could be more than two orders of magnitude faster than long crack growth at similar ΔK values, particularly when close to the long crack growth threshold, which was 8.7 MPa$\sqrt{m}$ for R = 0.33 (Brown and Hicks 1983).

Hicks and Brown (1982) also observed that although the short cracks arrested (briefly) at grain and colony boundaries, they did not slow down before arresting.

More detail about crystallographic crack growth in IMI 685 was obtained from extremely coarse-grained IMI 685 that had been β heat-treated for 7 days followed by slow cooling (Bache et al. 1998; Wilson et al. 1999). The prior β grain size was about 30 mm and the colony sizes ranged from 0.1 to 5 mm. Thin (2 mm) coupons containing 0.25 mm edge notches were fatigued to through-crack lengths of about 4 mm. During the first 0.5 mm of cracking the growth rates depended strongly on the crystallographic orientation of the initially cracking colony with respect to the principal stress axis. Once the cracks grew beyond the initial colony, at about 1–1.5 mm *total* length, the growth rates gradually converged to similar values.

5.2.2 Effects of Microstructure: Fully Lamellar

Three trends have been reported or proposed for the effects of fully lamellar microstructural variations on short/small fatigue crack growth over similar ranges of ΔK:

- Finer prior β grain size reduces the crack growth rates (Hastings et al. 1987). This conclusion was obtained from a crude comparison of small crack growth rate data for IMI 685, grain size ~ 5 mm (Brown and Hicks 1983) and Ti65S, grain size ~ 1.5 mm (Hastings et al. 1987). A more representative comparison is given in Fig. 5.4, which shows the data envelopes for microstructurally short crack growth across aligned α platelets. There is indeed a trend of lower crack growth rates for the finer-grained material, but there is also considerable overlap.
- A "basketweave"[2] microstructure results in lower crack growth rates than colonies of aligned α platelets (Hastings et al. 1987). This conclusion was based on the data in Fig. 5.5, notably a comparison of the data for crack sizes within the colony size range of the aligned microstructure. The data scatter makes this conclusion dubious, and it is also not supported by long/large fatigue crack growth rate data, see Sect. 6.3.
- A finer lamellar microstructure results in lower crack growth rates than a coarse lamellar microstructure (Wagner and Lütjering 1987). This conclusion was based on the data in Fig. 5.6. The data scatter makes this conclusion also somewhat dubious, but the highest overall growth rates do occur in the coarse lamellar material.

---

[2] "Basketweave" microstructures in titanium alloys consist of fine Widmanstätten configurations of intersecting α platelets within the prior β grains.

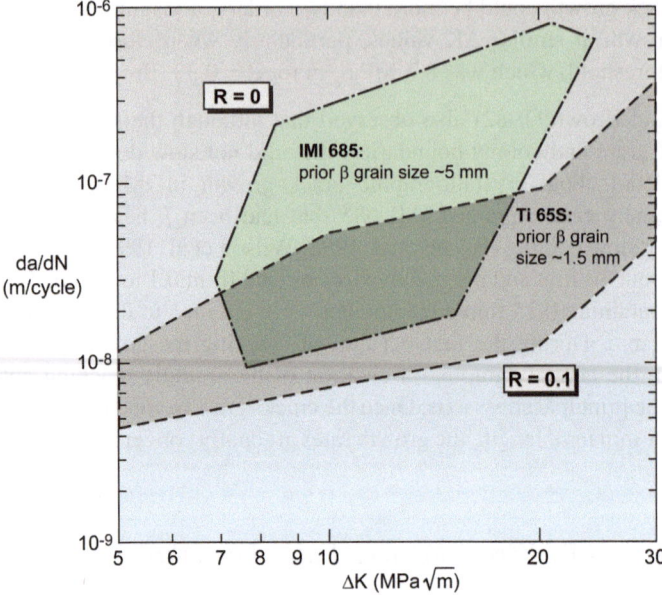

Fig. 5.4 Microstructurally short fatigue crack growth across aligned platelets in IMI 685 and Ti65S. After Brown and Hicks (1983) and Hastings et al. (1987)

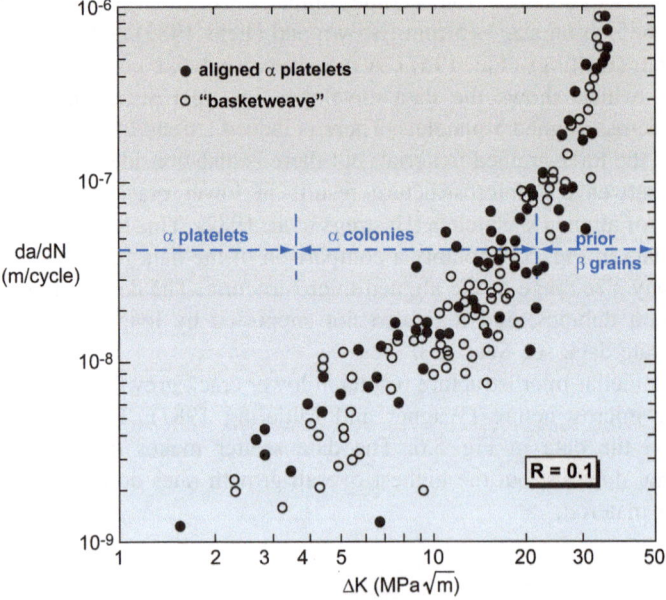

Fig. 5.5 Microstructurally short fatigue crack growth in Ti65S. After Hastings et al. (1987)

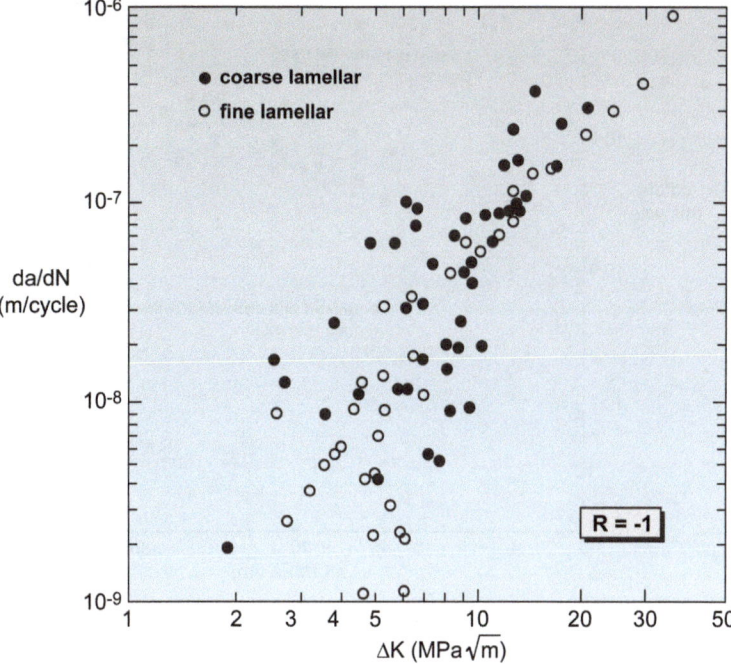

Fig. 5.6 Microstructurally short fatigue crack growth in Ti-6Al-4V. After Wagner and Lütjering (1987)

5.2.3 Effects of Microstructure: Different Types of Microstructures

Three trends have been reported or proposed for the effects of different types of microstructures on short/small fatigue crack growth over similar ranges of ΔK:

- Equiaxed primary α microstructures result in lower crack growth rates than coarse lamellar microstructures (Hicks and Brown 1984; Wagner and Lütjering 1987). This conclusion was obtained from the data shown in Figs. 5.7 and 5.8. The data scatter in Fig. 5.8 makes this conclusion dubious for fatigue crack growth at R = -1.
- Under reversed fatigue stressing (R = −1) equiaxed primary α microstructures result in lower crack growth rates than duplex and fine lamellar microstructures at ΔK values less than 5 and 10 MPa$\sqrt{}$m, respectively (Wagner and Lütjering 1987). This conclusion was based on the data shown in Figs. 5.6, 5.8 and 5.9. The data scatter in these figures makes this conclusion dubious.
- Duplex microstructures result in lower crack growth rates than coarse or fine lamellar microstructures (Wagner and Lütjering 1987; Dowson et al. 1992; Lütjering et al. 1993). This conclusion was based mainly on data shown in Figs. 5.6, 5.9 and 5.10. Figure 5.9 partially supports this conclusion, namely for

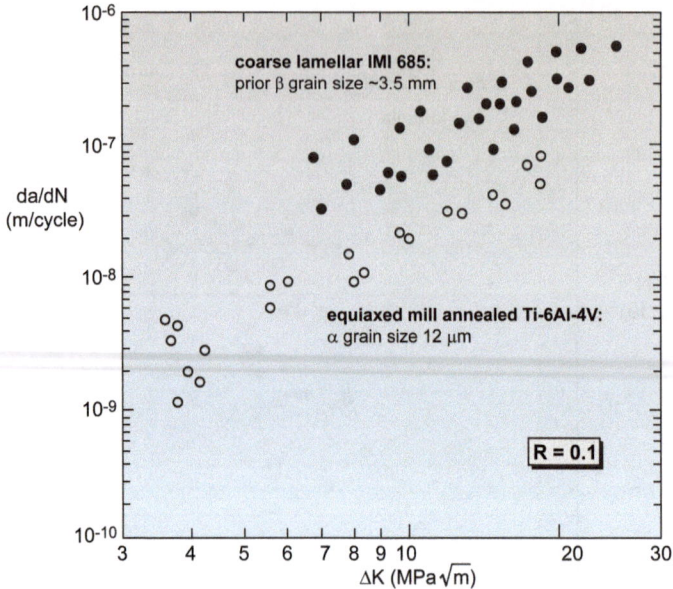

Fig. 5.7 Microstructurally short (IMI 685) and mechanically short (Ti-6Al-4V) fatigue crack growth comparisons (Hicks and Brown 1984)

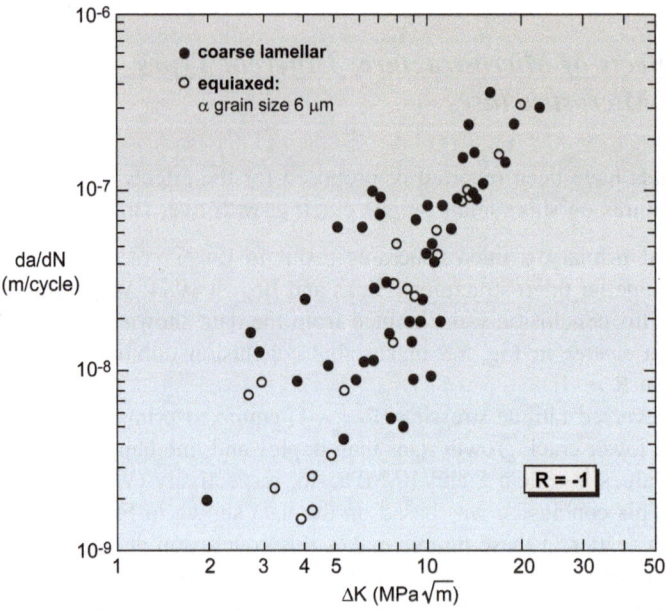

Fig. 5.8 Microstructurally short (coarse lamellar) and mechanically short (equiaxed) fatigue crack growth in Ti-6Al-4V. After Wagner and Lütjering (1987)

ΔK values *above* 10 MPa$\sqrt{}$m. However, Fig. 5.10 shows an opposing trend: generally lower crack growth rates occur in duplex microstructures only for ΔK values *below* 10 MPa$\sqrt{}$m.

The discrepancies between the reported or proposed trends and what Figs. 5.6, 5.8–5.10 indicate are due to data interpretation. Wagner and Lütjering reduced their data plots to "best fit" lines, thereby discounting the data scatter. Dowson et al. did this also for IMI 685 and Ti65S.

Wagner and Lütjering (1987) and Dowson et al. (1992) provided similar explanations of the trends derived from their "best fit" lines. They concluded that finer grain sizes and phase dimensions increase the number of microstructural barriers and improve the resistance to short/small crack growth. This explanation certainly has merit, but the data scatter shown in Figs. 5.6, 5.8–5.10 indicates that other factors should be considered. One likely possibility is the effect of local crystallographic orientation on the nucleation and early growth of cracks, already discussed in Sect. 5.2.1, and also mentioned by Brown and Taylor (1984) and Bache (1999).

5.3 Summary

As mentioned at the beginning of Sect. 5.1, short/small fatigue crack growth is a complex subject, since a variety of factors can affect the crack behaviour. In particular, the microstructures of titanium alloys can strongly influence early crack growth, with some clear trends and other not-so-clear trends, see Sect. 5.2.

Many, or most, aerospace alloys have average grain sizes less than about 50 μm. In view of the criteria in Table 5.1, this approximate grain size limit implies that microstructurally short/small fatigue crack growth behaviour should not persist beyond 0.5 mm. However, β processed and β heat-treated titanium alloys can have rather coarse microstructures. For example, the β annealed Ti-6Al-4V ELI plate that provides the motivation for the present publication has a prior β grain size averaging about 1.2 mm. For such coarse microstructures short/small fatigue crack growth behaviour may be expected to persist to crack sizes of several millimetres, owing to the strong influences of the local crystallographic orientations of grains and colonies of aligned α platelets. These influences are also responsible for wide variations in crack growth rates.

The persistence and variability of short/small fatigue crack growth behaviour in coarse-grained titanium alloys demonstrate that it must be investigated as part of a comprehensive study of fatigue life assessment methods for β annealed Ti-6Al-4V ELI plate.

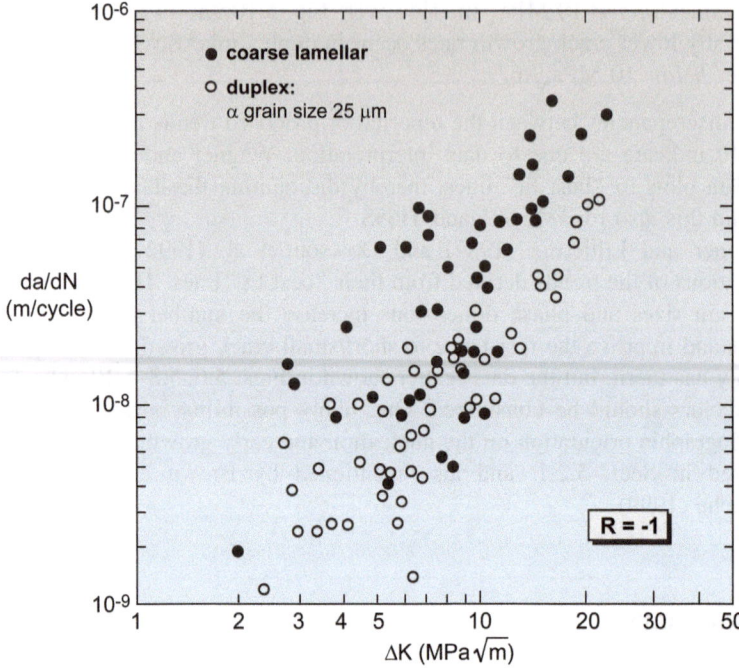

Fig. 5.9 Microstructurally short (coarse lamellar) and mechanically short (duplex) fatigue crack growth in Ti-6Al-4V. After Wagner and Lütjering (1987)

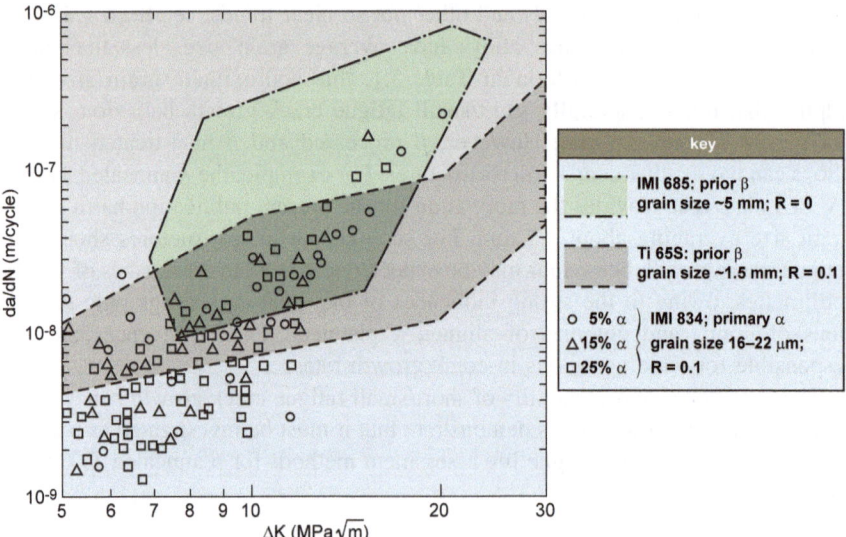

Fig. 5.10 Microstructurally short (IMI 685 and Ti65S) and mechanically short (IMI 834) fatigue crack growth comparisons. After Brown and Hicks (1983), Hastings et al. (1987) and Dowson et al. (1992)

References

R.F.W. Anstee, An assessment of the importance of small crack growth to aircraft design, in *Behaviour of Short Cracks in Aircraft Components*, AGARD Conference Proceedings No. 328, Advisory Group for Aerospace Research and Development, Neuilly-sur-Seine, France (1983), pp. 3-1–3-9

R.F.W. Anstee, P.R. Edwards, A review of crack growth threshold and crack propagation rates at short crack lengths, in *Some Considerations on Short Crack Growth Behaviour in Aircraft Structures*, AGARD Report No. 696, Advisory Group for Aerospace Research and Development, Neuilly-sur-Seine, France (1983), pp. 2-1–2-12

M.R. Bache, Microstructural influences on fatigue crack growth in the near alpha titanium alloy Timetal 834, in *Small Fatigue Cracks: Mechanics, Mechanisms and Applications*, ed. by K.S. Ravichandran, R.O. Ritchie, Y. Murakami (Elsevier Science Ltd, Oxford, 1999), pp. 179–186

M.R. Bache, W.J. Evans, V. Randle, R.J. Wilson, Characterization of mechanical anisotropy in titanium alloys. Mater. Sci. Eng. A **A257**, 139–144 (1998)

P.J. Bania, L.R. Bidwell, J.A. Hall, D. Eylon, A.K. Chakrabarti, Fracture—microstructure relationships in titanium alloys, in *Titanium and Titanium Alloys, Scientific and Technological Aspects*, vol. 1, ed. by J.C. Williams, A.F. Belov (Plenum Press, New York, 1982), pp. 663–677

C.W. Brown, M.A. Hicks, A study of short fatigue crack growth behaviour in titanium alloy IMI 685. Fatigue Eng. Mater. Struct. **6**, 67–76 (1983)

C.W. Brown, D. Taylor, The effects of texture and grain size on the short fatigue crack growth rates in Ti-6Al-4V, in *Fatigue Crack Growth Threshold Concepts*, ed. by D.L. Davidson, S. Suresh (The Metallurgical Society of AIME, Warrendale, 1984), pp. 433–446

J.E. Coyne, The beta forging of titanium alloys, in *The Science, Technology and Application of Titanium*, ed. by R.I. Jaffee, N.E. Promisel (Pergamon Press, London, 1970), pp. 97–110

E.R. De los Rios, H.J. Mohamed, K.J. Miller, A micromechanics analysis for short fatigue crack-growth. Fatigue Fract. Eng. Mater. Struct. **8**, 49–63 (1985)

X. Demulsant, J. Mendez, Microstructural effects on small fatigue crack initiation and growth in Ti6Al4V alloys. Fatigue Fract. Eng. Mater. Struct. **18**, 1483–1497 (1995)

M.J. Donachie, Jr., Introduction to titanium and titanium alloys, in *Titanium and Titanium Alloys Source Book*, ed. by M.J. Donachie Jr. (ASM International, Metals Park, 1982), p. 3

M.J. Donachie Jr., *Titanium: A Technical Guide*, 2nd edn. (ASM International, Materials Park, 2000), pp. 34–37

A.L. Dowson, A.C. Hollis, C.J. Beevers, The effect of the alpha-phase volume fraction and stress ratio on the fatigue crack growth characteristics of the near-alpha IMI 834 Ti alloy. Int. J. Fatigue **14**, 262–270 (1992)

W.J. Evans, Microstructure and the development of fatigue cracks at notches. Mater. Sci. Eng. A **A263**, 160–175 (1999)

T.E. Green, C.D.T. Minton, The effect of beta processing on properties of titanium alloys, in *The Science, Technology and Application of Titanium*, ed. by R.I. Jaffee, N.E. Promisel (Pergamon Press, London, 1970), pp. 111–119

P.J. Hastings, M.A. Hicks, J.E. King, The effect of α-platelet morphology and β-grain size on the initiation and growth of short fatigue cracks in Ti65S, in *Fatigue '87*, ed. by R.O. Ritchie, E.A. Starke Jr. (Engineering Materials Advisory Services Ltd, Warley, 1987), pp. 251–259

M.A. Hicks, C.W. Brown, Short fatigue crack growth in planar slip materials. Int. J. Fatigue **4**, 167–169 (1982)

M.A. Hicks, C.W. Brown, A comparison of short crack growth behaviour in engineering alloys, in *Fatigue 84*, ed. by C.J. Beevers (Engineering Materials Advisory Services Ltd, Warley, 1984), pp. 1337–1347

M.A. Hicks, C. Howland, C.W. Brown, Effect of load ratio and peak stress on short fatigue crack growth in β-processed titanium alloys, in *The Metallurgy of Light Alloys*, ed. by R.J. Taunt, P.J. Gregson (The Institution of Metallurgists, London, 1983), pp. 252–259

M.N. James, J.F. Knott, Aspects of small crack growth, in *Damage Tolerance Concepts for Critical Engine Components*, AGARD Conference Proceedings No. 393, Advisory Group for Aerospace Research and Development, Neuilly-sur-Seine, France (1985), pp. 10-1–10-12

H. Kitagawa, S. Takahashi, Applicability of fracture mechanics to very small cracks or cracks in the early stage, in *Proceedings of the Second International Conference on the Mechanical Behaviour of Materials (ICM-II)* (American Society for Metals, Metals Park, 1976), pp. 627–631

G. Lütjering, D. Helm, M. Däubler, Influence of microstructure on fatigue properties of the new titanium alloy IMI 834, in *Fatigue '93*, vol. 1, ed. by J.-P. Bailon, J.I. Dickson (Engineering Materials Advisory Services Ltd, Warley, 1993), pp. 165–170

R.C. McClung, K.S. Chan, S.J. Hudak, Jr., D.L. Davidson, Behavior of small fatigue cracks, in *ASM Handbook, Fatigue and Fracture*, vol. 19, ed. by S.R. Lampman et al. (ASM International, Materials Park, 1996), pp. 153–158

R.O. Ritchie, S. Suresh, Mechanics and physics of the growth of small cracks, in *Behaviour of Short Cracks in Aircraft Components*, AGARD Conference Proceedings No. 328, Advisory Group for Aerospace Research and Development, Neuilly-sur-Seine, France (1983), pp. 1-1–1-14

J. Schijve, Significance of fatigue cracks in micro-range and macro-range, in *Fatigue Crack Propagation, ASTM STP 415*, ed. by J.C. McMillan, R.M.N. Pelloux (American Society for Testing and Materials, Philadelphia, 1967), pp. 415–459

S. Suresh, R.O. Ritchie, The propagation of short fatigue cracks. Int. Metall. Rev. **29**, 445–476 (1984)

D. Taylor, Fatigue of short cracks: the limitations of fracture mechanics, in *The Behaviour of Short Fatigue Cracks*, ed. by K.J. Miller, E.R. de Los Rios (Mechanical Engineering Publications, London, 1986), pp. 479–490

G. Terlinde, T. Witulski, G. Fischer, Forging of titanium, in *Titanium and Titanium Alloys, Fundamentals and Applications*, ed. by C. Leyens, M. Peters (Wiley-VCH GmbH & Co. KGaA, Weinheim, 2003), pp. 289–304

L. Wagner, Fatigue life behavior, in *ASM Handbook Volume 19 Fatigue and Fracture, Second Printing*, ed. by S.R. Lampman et al. (ASM International, Materials Park, 1997), pp. 837–845

L. Wagner, G. Lütjering, Microstructural influence on propagation behavior of short cracks in an $(\alpha + \beta)$ Ti alloy. Zeitschrift Für Metallkunde **78**, 369–375 (1987)

R.J. Wilson, M.R. Bache, W.J. Evans, Crystallographic orientation and short fatigue crack propagation in a titanium alloy, in *Small Fatigue Cracks: Mechanics, Mechanisms and Applications*, ed. by K.S. Ravichandran, R.O. Ritchie, Y. Murakami (Elsevier Science Ltd, Oxford, 1999), pp. 199–206

Chapter 6
Long/Large Fatigue Crack Growth

6.1 Introduction

Long/large fatigue crack growth under constant amplitude loading can be considered in terms of the three regions shown in Fig. 6.1. In region I there is a threshold value, ΔK_{th}, below which cracks do not propagate. Above this value the crack growth rate increases relatively rapidly with increasing ΔK. In region II there is a more or less linear log–log relation between da/dN and ΔK. In region III the crack growth rate curve rises rapidly towards final failure.

Only regions I and II will be discussed in this report. For titanium alloys region II is often characterized by bilinear log–log relations between da/dN and ΔK (Yoder et al. 1977a, 1978, 1979, 1984; Gross et al. 1988; Wanhill et al. 1989; Saxena and Malakondaiah 1989; Ravichandran 1991; Wanhill and Looije 1993; Saxena and Radhakrishnan 1998; Wang and Müller 1998).

6.2 Fatigue Thresholds

6.2.1 Fully Lamellar Microstructures

Figure 6.2 shows ΔK_{th} values for β annealed Ti-6Al-4V as functions of aligned α colony size and R. Although the data are limited, Fig. 6.2a indicates that ΔK_{th} is fairly independent of α colony size. Ravichandran (1991) provided a detailed explanation of this result. He concluded that there was a change in the microstructural units controlling crack growth and threshold. For fast-cooled fine lamellar microstructures ("basketweave") the controlling microstructural unit (CMU) is the colony size; for relatively slow-cooled coarse lamellar microstructures (aligned α platelets) the CMU is the α platelet thickness.

R. Wanhill and S. Barter, *Fatigue of Beta Processed and Beta Heat-treated Titanium Alloys*, SpringerBriefs in Applied Sciences and Technology, DOI: 10.1007/978-94-007-2524-9_6, © The Author(s) 2012

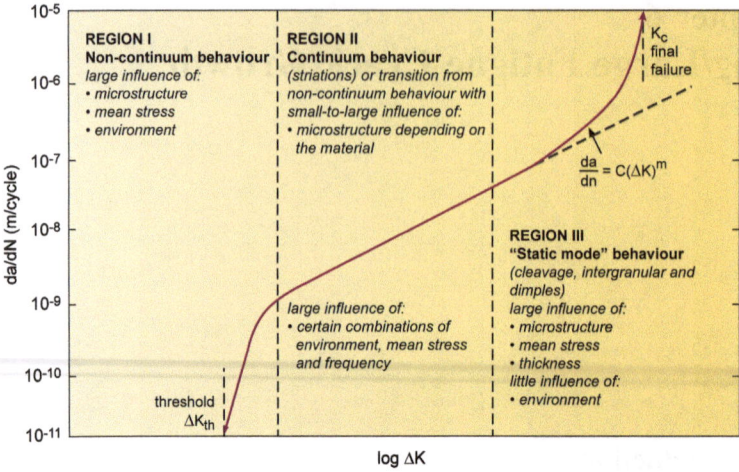

Fig. 6.1 Characteristics of the long/large fatigue crack growth rate curve log da/dN versus log ΔK (Ewalds and Wanhill 1984)

Fig. 6.2 Dependence of β annealed Ti-6Al-4V ΔK_{th} on **a** α colony size and **b** stress ratio, R. Solid symbol (•) data from Irving and Beevers (1974), Lütjering and Wagner (1988) and Ravichandran (1991); open symbol (o) data from NLR tests (2008)

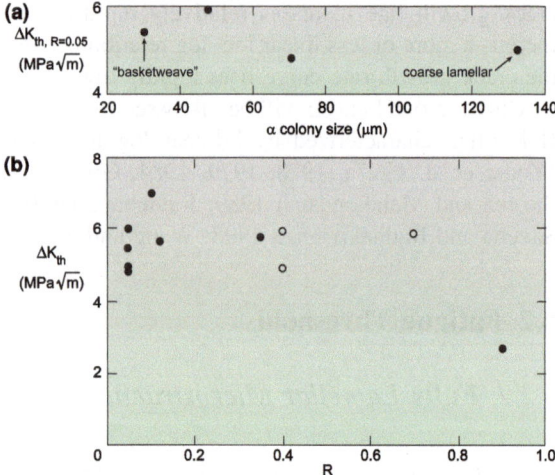

Figure 6.3 shows ΔK_{th} values for β processed and annealed IMI 685 (Hicks et al. 1983). The values for the coarsest (α colony size 0.21 mm) and finest ("basketweave") microstructures provide support for ΔK_{th} being fairly independent of α colony size. However, some of the data for the intermediate microstructure (α colony size 0.11 mm) are significantly lower.

Both Figs. 6.2b and 6.3 show a trend of more or less continuously decreasing ΔK_{th} with increasing R. This agrees with a general trend for titanium alloys with different types of microstructures (Chan 2004).

Fig. 6.3 Dependence of β processed and annealed IMI 685 ΔK_{th} on stress ratio, R (Hicks et al. 1983)

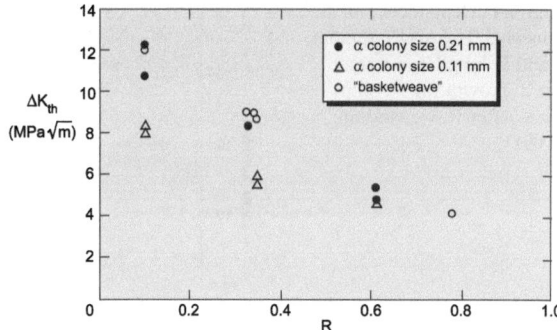

Table 6.1 Estimated fatigue thresholds for Ti-6Al-4V with different microstructures

R	Microstructure	ΔK_{th} (MPa$\sqrt{m}$)
−1	Coarse lamellar	7.2
	Fine lamellar	5.2
	Duplex	7.2
	Equiaxed primary α	4.6
0.35	Coarse lamellar	5.8
	Duplex	3.4
	Equiaxed primary α	3.1

6.2.2 Different Types of Microstructures

Table 6.1 gives estimated ΔK_{th} values for different Ti-6Al-4V microstructures. The estimates were obtained by extrapolation of near-threshold crack growth data (Irving and Beevers 1974; Wagner and Lütjering 1987) down to a crack growth rate of 10^{-10} m/cycle. Although limited in number, the estimates are for widely different R and show that coarse lamellar microstructures had the highest ΔK_{th} values, duplex microstructures were intermediate, and equiaxed primary α microstructures (usually the mill annealed condition) had the lowest ΔK_{th} values.

6.3 Regions I and II Fatigue Crack Growth in Fully Lamellar Microstructures

Figures 6.4 and 6.5 compare regions I and II long/large fatigue crack growth in coarse lamellar and "basketweave" β annealed Ti-6Al-4V and β processed and annealed IMI 685. For both alloys the near-threshold crack growth rates were very similar at the same or similar R values. These results are important because they imply that near-threshold long/large fatigue crack growth in β processed and β heat-treated titanium alloys is insensitive to variations in α colony size. This adds further support to Ravichandran's hypothesis that a change in the CMU is responsible for the similar crack growth behaviour, see Sect. 6.2.1.

Fig. 6.4 Dependence of β annealed Ti-6Al-4V regions I and II long/large fatigue crack growth on α colony size. After Ravichandran (1991)

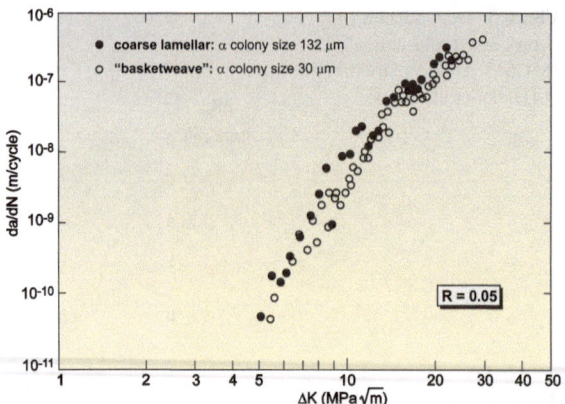

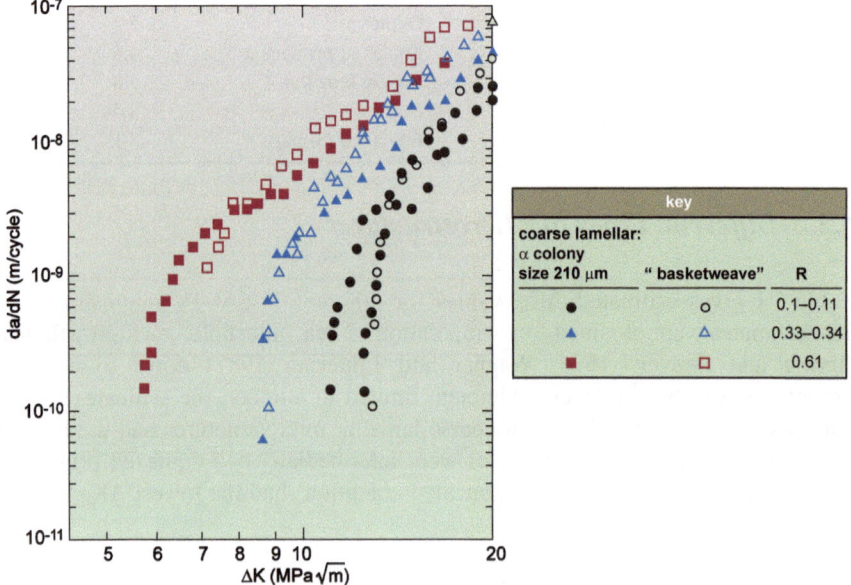

Fig. 6.5 Dependence of β processed and annealed IMI 685 regions I and II long/large fatigue crack growth on final microstructure (coarse lamellar and "basketweave") and stress ratio, R. After Hicks et al. (1983)

6.4 Regions I and II Fatigue Crack Growth in Different Types of Microstructures

Figures 6.6, 6.7 and 6.8 compare regions I and II long/large fatigue crack growth in different microstructures of Ti-6Al-4V and another α–β alloy, Ti-6Al-4Mo-2Zr-0.2Si. The coarse and fine lamellar microstructures tend to result in the lowest

Fig. 6.6 Dependence of
Ti-6Al-4V regions I and II
long/large fatigue crack
growth on microstructure.
After Wagner and Lütjering
(1987)

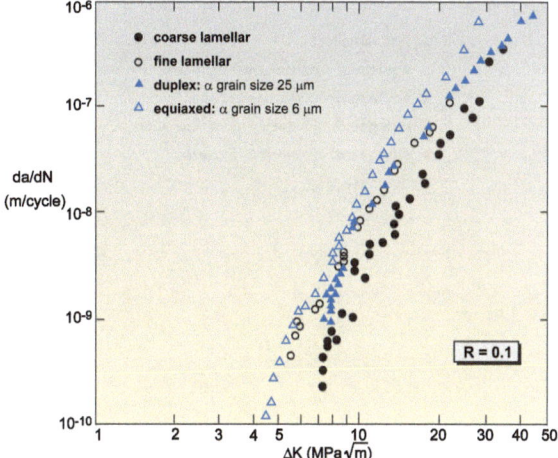

Fig. 6.7 Dependence of
Ti-6Al-4V regions I and II
long/large fatigue crack
growth on microstructure
(Irving and Beevers 1974).
STA = $(\alpha + \beta)$ Solution
Treated and Aged

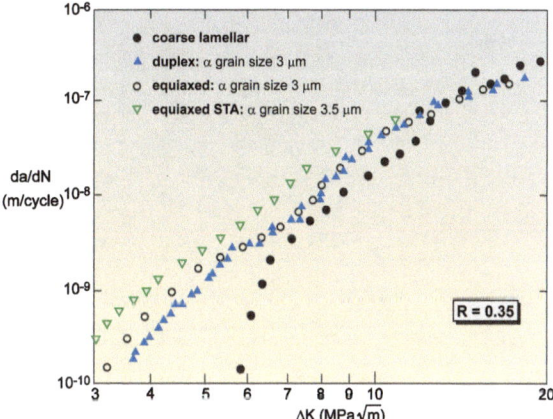

crack growth rates until da/dN values of about 10^{-7} m/cycle are exceeded. The crack growth rate curves for the other types of microstructures show varying amounts of overlap.

The lower fatigue crack growth rates in fully lamellar microstructures have been attributed to extensive crack branching and deflection (Yoder et al. 1976, 1977a, b; Eylon and Bania 1978; Eylon 1979; Bania et al. 1982) and enhanced roughness-induced fatigue crack closure (Halliday and Beevers 1981; Hicks et al. 1983; Saxena and Radhakrishnan 1998). Both these characteristics lower the fatigue crack driving force (Suresh and Ritchie 1982; Suresh 1983, 1985).

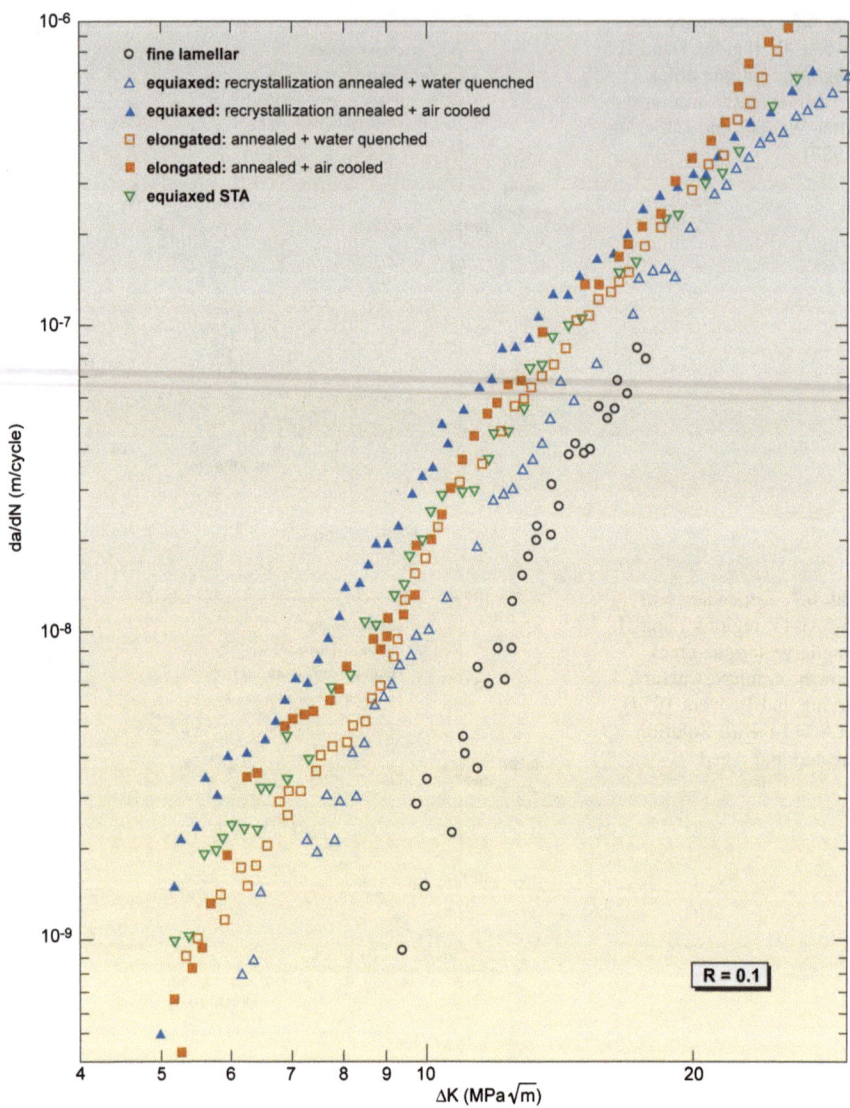

Fig. 6.8 Dependence of Ti-6Al-4Mo-2Zr-0.2Si regions I and II long/large fatigue crack growth on microstructure (Saxena and Radhakrishnan 1998). STA = $(\alpha + \beta)$ Solution Treated and Aged

6.5 Region II Bilinear log da/dN–log ΔK Fatigue Crack Growth

As mentioned in Sect. 6.1, region II fatigue crack growth in titanium alloys is often characterized by bilinear log–log relations between da/dN and ΔK. Figure 6.9 gives examples for (a) a β annealed Ti-6Al-4V ELI plate and (b) several alloys in

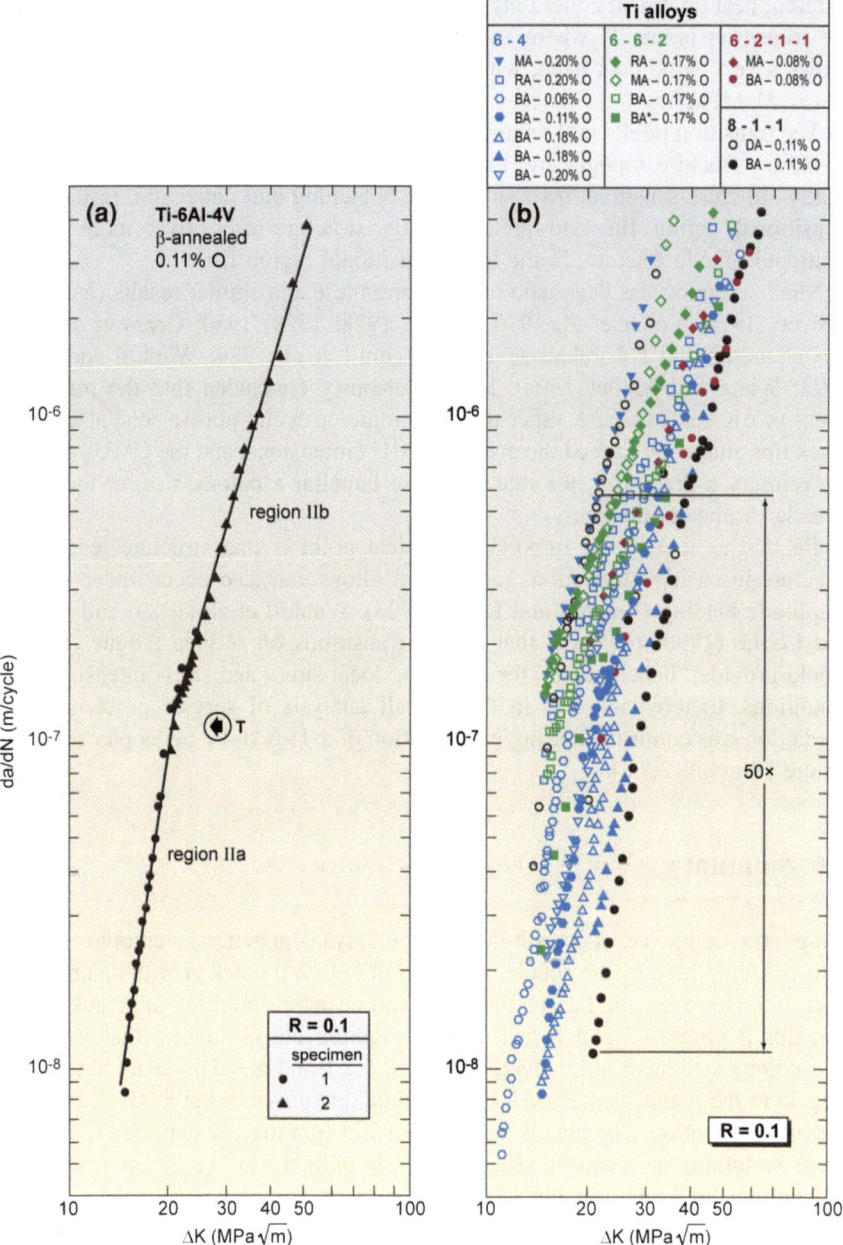

Fig. 6.9 Examples of region II fatigue crack growth for **a** a β annealed Ti-6Al-4V ELI plate and **b** several titanium alloys in different heat treatment conditions (Yoder et al. 1978, 1979)

different heat treatment conditions (Yoder et al. 1978, 1979). Figure 6.9a indicates the transition point, T, where the bilinear relation changes slope. Figure 6.9b shows a very wide data band, with a 50-fold difference in crack growth rates at $\Delta K = 21$ MPa$\sqrt{}$m.

The transition points in the bilinear plots (T in Fig. 6.9a) correspond to changes in fatigue fracture topography. For all these alloys crack growth was predominantly structure-sensitive, involving crack branching and deflection, in the hypo-transitional region IIa; and predominantly structure-insensitive, more or less continuum-mode fracture, in the hypertransitional region IIb.

Much attention has been paid to analysing these and similar results (Irving and Beevers 1974; Yoder et al. 1976, 1977b, 1978, 1979, 1980; Gross et al. 1988; Ravichandran and Dwarakadasa 1989; Wanhill et al. 1989; Wanhill and Looije 1993; Wang and Müller 1998). It was variously concluded that the transitions occur at ΔK values where either the monotonic or cyclic plastic zone sizes at the crack tips attain and exceed the average CMU dimensions; and the CMUs could be the primary α grain size, the fine or coarse lamellar α colony size, or the coarse lamellar α platelet thickness.

Be that as it may, an important practical point is that structure-sensitive to structure-insensitive transitions in titanium alloys can also occur under variable amplitude loading (Wanhill and Looije 1993). Wanhill et al. (1989) and Wanhill and Looije (1993) predicted that visible transitions on service fatigue fractures would provide "benchmarks" for analysing local stress and stress intensity factor conditions, thereby assisting in the overall analysis of service problems. This prediction was confirmed during investigation of a Ti-6Al-4V helicopter rotor hub failure (Wanhill 2003).

6.6 Summary

Long/large fatigue crack growth in titanium alloys is in general a complex subject. This is exemplified by the detailed analyses of region II crack growth mentioned in Sect. 6.5. However, the fatigue threshold and near-threshold fatigue crack growth data for β processed and β heat-treated titanium alloys suggest that variations in α colony size have little effect, see Sects. 6.2.1 and 6.3. This is important with respect to the β annealed Ti-6Al-4V ELI plate that provides the motivation for the present publication. The plate's thickness of 125 mm may be expected to result in some variations in α colony size, but this is unlikely to significantly affect the fatigue threshold and near-threshold fatigue crack growth behaviour.

Another important aspect is the structure-sensitive to structure-insensitive transition during region II fatigue crack growth. The ΔK value for this transition may show some variation through the plate thickness. The fracture topography characteristics of this transition should be determined. As pointed out in Sect. 6.5, the transition can act as a "benchmark" for fracture mechanics analyses of service failures.

References

P.J. Bania, L.R. Bidwell, J.A. Hall, D. Eylon, A.K. Chakrabarti, Fracture—microstructure relationships in titanium alloys, in *Titanium and Titanium Alloys, Scientific and Technological Aspects*, vol. 1, ed. by J.C. Williams, A.F. Belov (Plenum Press, New York), pp. 663–677

K.S. Chan, Variability of large-crack fatigue-crack-growth thresholds in structural alloys. Metall. Mater. Trans. A **35A**, 3721–3735 (2004)

H.L. Ewalds, R.J.H. Wanhill, *Fracture Mechanics* (Edward Arnold (Publishers) Ltd and Delftse Uitgevers Maatschappij b.v., London, and Delft, 1984), pp. 172–173

D. Eylon, Faceted fracture in beta annealed titanium alloys. Metall. Trans. A **10A**, 311–317 (1979)

D. Eylon, P.J. Bania, Fatigue cracking characteristics of β-annealed large colony Ti-11 alloy. Metall. Trans. A **9A**, 1273–1279 (1978)

T.S. Gross, S. Bose, L. Zhong, Frictional effects on fatigue crack growth in β-annealed Ti-6Al-4V. Fatigue Fract. Eng. Mater. Struct. **11**, 179–187 (1988)

M.D. Halliday, C.J. Beevers, Some aspects of fatigue crack closure in two contrasting titanium alloys. J. Testing Eval. **9**, 195–201 (1981)

M.A. Hicks, R.H. Jeal, C.J. Beevers, Slow fatigue crack growth and threshold behaviour in IMI 685. Fatigue Eng. Mater. Struct. **6**, 51–65 (1983)

P.E. Irving, C.J. Beevers, Microstructural influences on fatigue crack growth in Ti-6Al-4V. Mater. Sci. Eng. **14**, 229–238 (1974)

G. Lütjering, L. Wagner, Influence of texture on fatigue properties of titanium alloys, in *Directional Properties of Materials*, ed. by H.J. Bunge (DGM Informationsgesellschaft mbH, Oberursel, 1988), pp. 177–188

K.S. Ravichandran, Near threshold fatigue crack growth behavior of a titanium alloy: Ti-6Al-4V. Acta Metall. et Mater. **39**, 401–410 (1991)

K.S. Ravichandran, E.S. Dwarakadasa, Fatigue crack growth transitions in Ti-6Al-4V alloy. Scripta METALLURGICA **23**, 1685–1690 (1989)

V.K. Saxena, G. Malakondaiah, Effect of heat treatment on fatigue crack growth in Ti-6Al-3.5Mo-1.9Zr-0.23Si alloy. Int. J. Fatigue **11**, 423–428 (1989)

V.K. Saxena, V.M. Radhakrishnan, Effect of phase morphology on fatigue crack growth behavior of α-β titanium alloy—a crack closure rationale. Metall. Mater. Trans. A **29A**, 245–261 (1998)

S. Suresh, Crack deflection: implications for the growth of long and short fatigue cracks. Metall. Trans. A **14A**, 2375–2385 (1983)

S. Suresh, Fatigue crack deflection and fracture surface contact: micromechanical models. Metall. Trans. A **16A**, 249–259 (1985)

S. Suresh, R.O. Ritchie, A geometrical model for fatigue crack closure induced by fracture surface morphology. Metall. Trans. A **13A**, 1627–1631 (1982)

L. Wagner, G. Lütjering, Microstructural influence on propagation behavior of short cracks in an (α + β) Ti alloy. Zeitschrift für Metallkunde **78**, 369–375 (1987)

S.-H. Wang, C. Müller, A study on the change of fatigue fracture mode in two titanium alloys. Fatigue Fract. Eng. Mater. Struct. **21**, 1077–1087 (1998)

R.J.H. Wanhill, Material-based failure analysis of a helicopter rotor hub. Practical Fail. Anal. **3**, 59–69 (2003)

R.J.H. Wanhill, R. Galatolo, C.E.W. Looije, Fractographic and microstructural analysis of fatigue crack growth in a Ti-6Al-4V fan disc forging. Int. J. Fatigue **11**, 407–416 (1989)

R.J.H. Wanhill, C.E.W. Looije, Fractographic and microstructural analysis of fatigue crack growth in Ti-6Al-4V fan disc forgings, in *AGARD Engine Disc Cooperative Test Programme*, AGARD Report 766 (Addendum), Advisory Group for Aerospace Research and Development, Neuilly-sur-Seine, France (1993), pp. 2-1–2-40

G.R. Yoder, L.A. Cooley, T.W. Crooker, A micromechanistic interpretation of cyclic crack-growth behavior in a beta-annealed Ti-6Al-4V alloy, NRL Report 8048, (Naval Research Laboratory, Washington, 1976)

G.R. Yoder, L.A. Cooley, T.W. Crooker, Enhancement of fatigue crack growth and fracture resistance in Ti-6Al-4V and Ti-6Al-6V–2Sn through microstructural modification. Trans. ASME, J. Eng. Mater. Technol. **99**, 313–318 (1977a)

G.R. Yoder, L.A. Cooley, T.W. Crooker, Observations on microstructurally sensitive fatigue crack growth in a Widmanstätten Ti-6Al-4V alloy. Metall. Trans. A **8A**, 1737–1743 (1977b)

G.R. Yoder, L.A. Cooley, T.W. Crooker, Fatigue crack propagation resistance of beta-annealed Ti-6Al-4V alloys of differing interstitial oxygen contents. Metall. Trans. A **9A**, 1413–1420 (1978)

G.R. Yoder, L.A. Cooley, T.W. Crooker, 50-fold difference in region-II fatigue crack propagation resistance of titanium alloys: a grain size effect. Trans. ASME, J. Eng. Mater. Technol. **101**, 86–90 (1979)

G.R. Yoder, L.A. Cooley, T.W. Crooker, Observations on the generality of the grain-size effect on fatigue crack growth in $\alpha + \beta$ titanium alloys, in *Titanium '80 Science and Technology*, ed. by H. Kimura, O. Izumi (The Metallurgical Society of AIME, Warrendale, 1980), pp. 1865–1873

G.R. Yoder, F.H. Froes, D. Eylon, Effect of microstructure, strength, and oxygen content on fatigue crack growth rate of Ti-4.5Al-5.0Mo-1.5Cr (CORONA 5). Metall. Trans. A **15A**, 183–197 (1984)

Chapter 7
Concluding Remarks

This publication is a review of most of the available literature on the fatigue properties of β annealed Ti-6Al-4V and titanium alloys with similar microstructures. The focus is on β processed and β heat-treated titanium alloys. This is because β annealed Ti-6Al-4V ELI plate has been selected for the main wing-carry-through bulkhead and other fatigue critical structures, including the vertical tail stubs, of advanced military aircraft currently intended to enter service with several Air Forces around the world, including the RAAF and RNLAF. However, some comparisons are made with alloys having different microstructures, in particular conventionally $(\alpha + \beta)$ processed and heat-treated Ti-6Al-4V.

The review has been necessarily limited to fatigue under constant amplitude loading. There appears to be no open-source literature on the fatigue behaviour of β processed and β heat-treated titanium alloys under variable amplitude loading, in particular the load histories representative for military airframe components. Also, nothing is known about the ability of constitutive and crack growth models to predict this behaviour.

This situation emphasizes the practical usefulness of the DSTO—NLR programme to develop damage tolerance and durability assessment methods for β annealed Ti-6Al-4V ELI plate.

R. Wanhill and S. Barter, *Fatigue of Beta Processed and Beta Heat-treated Titanium Alloys*, SpringerBriefs in Applied Sciences and Technology, DOI: 10.1007/978-94-007-2524-9_7, © The Author(s) 2012